Failure Analysis

Edited by Zheng-Ming Huang
and Sayed Hemeda

Published in London, United Kingdom

IntechOpen

Supporting open minds since 2005

Failure Analysis
http://dx.doi.org/10.5772/intechopen.75250
Edited by Zheng-Ming Huang and Sayed Hemeda

Contributors
Ali Aljaroudi, Qingshan Feng, Alireza Khalifeh, Zheng-Ming Huang, Yury Golovin, Dmitry Golovin, Alexander Tyurin, Alexander Samodurov, Alexander Divin, Kenneth Crawford, Dongfeng He

First published in London, United Kingdom, 2019 by IntechOpen
IntechOpen is the global imprint of INTECHOPEN LIMITED, registered in England and Wales, registration number: 11086078, 7th floor, 10 Lower Thames Street, London, EC3R 6AF, United Kingdom
Printed in Croatia

British Library Cataloguing-in-Publication Data
A catalogue record for this book is available from the British Library

Additional hard and PDF copies can be obtained from orders@intechopen.com

Failure Analysis
Edited by Zheng-Ming Huang and Sayed Hemeda
p. cm.
Print ISBN 978-1-83968-253-7
Online ISBN 978-1-83968-254-4
eBook (PDF) ISBN 978-1-83968-255-1

For EU product safety concerns:
IN TECH d.o.o., Prolaz Marije Krucifikse Kozulić 3, 51000 Rijeka, Croatia, info@intechopen.com or visit our website at intechopen.com.

Meet the editors

Dr. Zheng-Ming Huang is a professor at the School of Aerospace Engineering & Applied Mechanics, Tongji University, China. He is known for his unified elastic-plastic constitutive theory and the bridging model for composites. He has found that the homogenized stresses in the constituents of a composite obtained by any micromechanics theory must be converted into true values before the effective property, specifically failure and strength behavior, can be determined from the monolithic constituent properties. He has established a systematic theory to accomplish the conversion. Challenging issues such as when the interface debonding occurs between the constituents of a composite subjected to any load have been addressed using his theories. He is the author/co-author of more than 200 papers, 4 books, 7 book chapters, and 22 patents. One of his papers has received more than 5,000 citations in Web of Science.

Dr Sayed Hemeda is a Doctor of the Civil Engineering Department, Aristotle University of Thessaloniki, Greece. He currently occupies the position of Professor of Geotechnical Engineering and Architectural Preservation of Architectural Heritage, Conservation Department, Faculty of Archaeology, Cairo University, Egypt. He is also the vice manager of the Historic Buildings Conservation Center in Cairo University.
He was awarded the Cairo University's prize of scientific excellence in 2017 and the Cairo University's prize of encouragement in 2014. Furthermore, he was awarded the Cairo University's prize for the best Ph.D Thesis, 2009-2010 and was awarded the General Union of Arab Archaeologists prize for academic excellence.
He has published approximately 75 articles, including 21 books and has been cited approximately 120 times.
He has given over 38 invited lectures in 16 countries. His interests are primarily in geotechnical engineering for architectural heritage preservation, as well as engineering data analysis, including pattern recognition as applied to primarily analytical data from various sources including objects of cultural significance.
He is the Editor-In-Chief of the Journal of Geological Research, an Editorial Board Member of the Sustainable Civil Infrastructures book series published by Springer Nature, an Editorial Board Member at IntechOpen, an Editorial Board Member of the Progress of Electrical and Electronic Engineering in Singapore, an Editorial Board Member ofn the Geoscience Journal in Singapore and an Editorial Board Member of the Alexandria Engineering Journal.

Contents

Preface

Failure of a material or structure is one of the oldest topics that engineers and researchers have to deal with. Traditionally, failure is related to an ultimate load sustaining ability of the material or structure, as taught in an elementary Strength of Materials textbook. Nowadays, the definition of failure has been extended from the traditional narrow sense to a much broader sense. Any expected function of a material or structure that cannot be fulfilled is considered to have failed.

This book brings together a small collection of chapters but covers a broad range of failures. Following the Introductory chapter, the second chapter presents the accident of a pipeline for oil or gas transportation, which can be considered as, in the broadest sense, failure of the pipeline. Any leakage will cause the pipeline to cease its transportation function, although the structure of the pipeline may not be damaged from a traditional viewpoint. A cause theory for the pipeline failure is described in the chapter, and a probable model for statistical analysis for the accident is given. Fragility, anti-fragility, and integrity are used as indexes to indicate the state of accident, providing a new way to assess an accident, which is different from the traditional accident assessment.

The third chapter deals with a stress corrosion cracking (SCC) failure that occurs only in metals or alloys. Three factors are necessary for this kind of failure to take place. One is a susceptible material, the second is a corrosive environment, and the third is a tensile stress generated in the material. Design of a structure with the occurrence of a SCC is generally based on or checked with fracture mechanics methodology, and the measured fracture toughness of a material must incorporate the effect of the corrosive environments.

In the fourth chapter, the traditional and probably narrowest sense of failure is addressed. These failures are not all well understood. Composite failures present great challenges to engineers and academic researchers. This chapter focuses on the failure analysis of a unidirectional composite based on an innovative concept of true stresses. The internal stresses in its constituent fiber and matrix are determined using a micromechanics theory. It is noted that these stresses are homogenized quantities. They must be converted into true values before a failure detection can be made against the strengths of the constituents. Once a constituent has failed, the composite is considered to have attained a failure status. How to convert the homogenized stresses into true quantities is illustrated in the chapter.

While the first three chapters treat the failures of materials and structures with deterministic methods, a probabilistic technique for failure analysis is explained in the fifth chapter. In general, deterministic techniques ignore the variability and uncertainties of the variables in a failure analysis. Contrastingly, probabilistic techniques incorporate the variability and uncertainties in the analysis. In this chapter, the commonly used probabilistic failure analysis techniques and their mathematical derivations are presented. Examples to enhance the understanding of the concept of failure analysis are also shown.

The non-destructive testing (NDT) methods are presented in the sixth chapter to test fiber reinforced polymers (FRP) - concrete on reinforced concrete (RC) structures using the impulse excitation principle. A light mobile impact machine using the impact-echo NDT method with multiple hammer impacts across the length of an FRP (CFRP) plate bonded to a concrete structure generates specific frequencies and waveforms. The frequencies and waveforms are unique and a function of the FRP-concrete bond condition. Detecting changes in frequencies and waveforms indicates a change in bond condition, i.e. from a bonded to a de-bonded state. This chapter explains the NDT method used to evaluate FRP-concrete bond on RC bridges in the field, outlines the impulse excitation theory, and the mathematics of the signal analysis to measure the impact signal frequencies and waveforms. Examples of NDT testing on several bridges in Missouri are presented with test results.

The seventh chapter describes how electromagnetic methods undertake nondestructive evaluation of the steel reinforcing bar (rebar) in concrete. AC current was passed through an excitation coil and an AC magnetic field was produced. Then an eddy current was induced in the steel rebar. A detection coil was used to measure the magnetic field produced by the eddy current and the magnetization effect.

Finally, the eighth chapter presents the temperature diffusivity measurement and nondestructive testing using non-stationary local heating and IR thermography for the engineering failure analysis.

We would like to express our gratitude to the publisher, IntechOpen for the efforts in publishing this book and to the authors of the accepted manuscripts for their work and patience.

Zheng-Ming Huang, PhD
Professor,
School of Aerospace Engineering and Applied Mechanics,
Tongji University,
Shanghai, P. R. China

Dr Sayed Hemeda
Professor in Geotechnical Engineering
and Architectural Preservation of Historic Buildings,
Cairo University,
Giza, Egypt

Chapter 1

Introductory Chapter: Failures Analysis

Zheng-Ming Huang and Sayed Hemeda

1. Introduction

Failure of a material or structure represents a multidiscipline characteristic of it. There is no universally applicable definition on the failure. In a broad sense, failure represents a behavior at which the material or structure is no longer able to fulfill its function. In a narrow sense, however, failure is defined as that given in an elementary strength of materials textbook. Namely, the narrow sense failure of a material or structure is related to its load sustaining ability. When the stresses sustained by the material or structure exceed their ultimate values, which are generally the strengths of the material, the material or the structure is said to have attained a failure status. Thus, a failure criterion in which the critical parameters are the ultimate values must be provided to assess the failure.

Mechanical failures, as a result of improper design, corrosion, surface fracture, and other material defects, are described in the context of real world. Case studies involve steam generators, boiler tubes, gas turbine blades, welded structures, chemical conversion reactors, and more.

Nondestructive testing (NDT) methods are important tools to help us assure the safety, quality, and reliability of materials and structures. Finding defects, whether they are flaws or imperfections, during the manufacturing, and construction process or before, are the first step in any "postmortem" failure analysis.

The most common types of NDT methods are:

- Visual inspection
- Dye (liquid) penetrant
- Magnetic particle inspection
- Radiographic (X-ray) testing
- Eddy current
- Ultrasonics
- Leak detection
- Acoustic emission
- Infrared thermography

Each of these methods has its advantages and limitations, and often more than one technique is needed to identify the root cause and quantify the extent of the problem.

2. Highlight on failure analysis

The broadest sense failure is concerned with an accident of a pipeline used for oil or gas transportation. A leakage of the pipeline is a most often occurred accident, which will cause the pipeline to cease its transportation function and hence is defined as a failure of the pipeline. Based on the accident survey and management practice, a list of characteristics and quantitative descriptions of the pipeline accidents or failures are summarized in the book. A probable model from statistical analysis for the accident is provided. Fragility, anti-fragility, and integrity are used as indexes to describe the state of an accident.

A stress corrosion cracking represents a narrower sense of failures. Three factors are necessary for the occurrence of such a failure. They are a sensitized material, a corrosive environment, and a tensile stress applied on the material. A stress intensity based failure criterion has been widely used to detect the failure. However, the measured fracture toughness of a material must take the effects of corrosive environments into account, and a significant cutting down on the fracture toughness can be seen.

The failure of a unidirectional composite is a typical example of the narrowest failure cases. The composite is anisotropic and its failure behavior is generally different at a different point as well as along a different direction. This makes the analysis of composite failures a great challenge. There are essentially two kinds of approaches to a composite failure, that is, phenomenological and micromechanical. Hardly any a composite failure can be efficiently addressed only through a phenomenological approach, since it does not care for any stresses induced in the constituent fiber and matrix of the composite. For instance, an interface debonding between the fiber and matrix, which frequently initiates the composite failure, cannot be practically dealt with without knowing the stress states in the fiber and matrix.

While most failure analyses are performed with deterministic methods, there are probabilistic techniques to theoretically characterize a failure of a material or structure as well. In a deterministic method, the critical parameters such as strengths of the material or structure are provided, without considering any variability and uncertainty involved in the determination of these parameters. On the other hands, the variability and uncertainties have been incorporated in a probabilistic technique to quantify a material's or structure's failure. A probabilistic technique is described in the book to determine a material failure.

3. Non-destructive characterization

The non-destructive testing (NDT) methods to test FRP-concrete on RC structures use an impulse excitation principle. In an impact-echo NDT method, a light mobile impact machine with multiple hammer impacts across the length of an FRP (CFRP) plate bonded to a concrete structure generates specific frequencies and waveforms. The frequencies and waveforms are unique as a function of the FRP-concrete bond condition. Detecting changes in frequencies and waveforms indicate a change in bond condition, that is, from a bonded to a de-bonded state. The NDT method used to evaluate FRP-concrete bond on RC bridges in the field is explained.

The impulse excitation theory and the mathematics of a signal analysis to measure the impact signal frequencies and waveforms are highlighted. Examples of NDT testing on several bridges in Missouri are presented.

Electromagnetic methods have been developed to do nondestructive evaluation of the steel reinforcing bar (rebar) in concrete. AC current flew in an excitation coil and AC magnetic field is produced, and eddy current is induced in the steel rebar. A detection coil is then used to measure the magnetic field produced by the eddy current and the magnetization effect.

A temperature diffusivity measurement and nondestructive testing using non-stationary local heating and IR thermography for analysis of some engineering failures are also outlined in the book.

The coverage of failures is very broad. Hardly any a single book is able to describe failure mechanisms or to provide diagnoses or detections on failures occurred in all kinds of materials and structures. In fact, the existing literature on the failure related topics is extensive. Some of them can be recommended for the readers to obtain more information. They include monographs or edited books by Holm and Tomasz [1], Velázquez and Luis [2], Makhlouf and Aliofkhazraei [3], Greuter and Zima [4], and Hinton et al. [5], and review papers by Blandford [6], Orifici et al. [7], Ossai et al. [8], Liu and Zheng [9], Breitenstein and Sturm [10], Louis [11], Hellier [12], Hemeda and Pitilakis [13], Hemeda [14], Giovanni [15], and among others. Even so, we believe the present book is useful to the readers.

Author details

Zheng-Ming Huang[1*] and Sayed Hemeda[2]

1 School of Aerospace Engineering and Applied Mechanics, Tongji University, P.R. China

2 Geotechnical Engineering and Architectural Preservation of Historic Buildings, Cairo University, Egypt

*Address all correspondence to: huangzm@tongji.edu.cn

References

[1] Holm A, Tomasz S. Failure and Damage Analysis of Advanced Materials. Heidelberg: Springer; 2015

[2] Velázquez G, Luis J. Fractography and Failure Analysis. Heidelberg: Springer; 2018

[3] Makhlouf ASH, Aliofkhazraei M, editors. Handbook of Materials Failure Analysis with Case Studies from the Chemicals, Concrete and Power Industries. Butterworth-Heinemann: Elsevier; 2016

[4] Greuter E, Zima S. Internal Combustion Engine Failures and their Causes. Warrendale: SAE International; 2012

[5] Hinton MJ, Kaddour AS, Soden PD, editors. Failure Criteria in Fibre Reinforced Polymer Composites—The World-wide Failure Exercise. Elsevier; 2004

[6] Blandford GE. Review of progressive failure analyses for truss structures. Journal of Structural Engineering. 1997;**123**(2):122-129

[7] Orifici AC, Herszberg I, Thomson RS. Review of methodologies for composite material modelling incorporating failure. Composite Structures. 2008;**86**:194-210

[8] Ossai CI, Boswell B, Davies IJ. Pipeline failures in corrosive environments—A conceptual analysis of trends and effects. Engineering Failure Analysis. 2015;**53**:36-58

[9] Liu PF, Zheng JY. Recent developments on damage modeling and finite element analysis for composite laminates: A review. Materials and Design. 2010;**31**(8):3825-3834

[10] Breitenstein O, Sturm S. Lock-in thermography for analyzing solar cells and failure analysis in other electronic components. Quantitative InfraRed Thermography Journal. 2019:1-15. DOI: 10.1080/17686733.2018.1563349

[11] Louis C. Nondestructive Testing. Materials Park, OH, US: ASM International; 1995. ISBN: 978-0-87170-517-4

[12] Hellier C. Handbook of Nondestructive Evaluation. US: McGraw Hill companies; 2003. p. 1.1. ISBN: 978-0-07-028121-9

[13] Hemeda S, Pitilakis K. Geophysical Investigations at Cairo's Oldest, the Church of Abu Serga (St. Sergius), Cairo, Egypt. Research in Nondestructive Evaluation. 2017;**28**(3):123-149. DOI: 10.1080/09349847.2016.1143991

[14] Hemeda S. Geotechnical and geophysical investigation techniques in Ben Ezra Synagogue in Old Cairo area, Egypt. Heritage Science. 2019;7:23. DOI: 10.1186/s40494-019-0265-y

[15] Giovanni L. Nondestructive Testing for Archaeology and Cultural Heritage. Switzerland AG: Springer Nature; 2019. DOI: 10.1007/978-3-030-01899-3

Chapter 2

Pipeline Failure Cause Theory: A New Accident Characteristics, Quantification, and Cause Theory

Qingshan Feng

Abstract

Based on the accident research and management practices of oil and gas pipelines, the characteristics and the quantitative description of the accident/failure are set up. Several characteristics are summarized which clearly describe the essential prosperities of the accident. Fragility, anti-fragility, and integrity are used as an index to describe the state of accident, which provides a new way of evaluating and describing accident, different from the traditional accident assessment. The understanding and the evaluation of the nature of accident become clearer. Accident cause theory is the basic theory of cognition and prevention of failure. In this chapter, based on the analysis of characteristics and limitations of some accident cause theories, and comprehension of characteristics of failure and systematic statistics, a new systematic accident cause theory is proposed, named by analogy with "tree-type." This theory provides a systematical supplement of accident cause theories.

Keywords: accident, failure, pipeline, causing theory, fragility, anti-fragility, integrity

1. Introduction

As the most safe, environmentally friendly and energy-saving oil and gas transmission method, pipelines have been widely constructed around the world. Compared with other methods of transportation, its safety is the highest, but more serious accidents can also occur, causing the public and the government to pay close attention in recent years [1]. Therefore, a correct understanding of pipeline accidents, prediction, and prevention of pipeline accidents is the basis for ensuring the safe operation of pipelines [2]. However, research in this area is mostly concentrated in the fields of construction, processing, and manufacturing of traditional industries. The study of accident causes, prediction, and prevention for the pipeline industry is relatively rare, which severely restricts the perception of pipeline accidents and limits the pertinence of the prepared response measures. The lack of theoretical basis for integrity management with the main purpose of preventing accidents has weakened the effect of pipeline integrity management [3].

First, this chapter discusses the characteristics of accidents and quantitative description methods. It can help to understand accidents logically and avoid deviations or mistakes in the cognitive process of accidents. Then, based on the cognition of the characteristics of accident, the author introduced the concept of fragile assessment to quantify the state of the system. It provided a new accident assessment description model for the industry and made it possible to recognize the cause of the accident and more formulated [4].

Accident cause theory is the theoretical basis for recognizing accidents and preventing similar accidents from occurring again. The current accident causality theories have certain limitations; they cannot find full life cycle problems and achieve targeted prevention. With the development of technology, the basis for analyzing the causes of accidents has changed, for example, it is easier to obtain a lot of data. This chapter is based on the ability to obtain a large amount of data, the cognition of accident characteristics, the analysis conclusions and statistics of a large number of accident causes, etc., proposes a theory of systematically analyzing the causes of accidents. This theory analogizes the life of the tree to describe this set of the accident cause theory. It is a new set of system analysis methods to the analysis of accident causes [5].

In order to predict and prevent accidents, it needs to know the cause of the accident. The era of big data makes this possible. In the past few years, big data technology has developed rapidly and has been widely used. Based on in-line inspection data or/and global positioning system (GPS) data, it is easy to setup the data model and make all data in one big database. The big data makes "samples = all"; it has the conditions of system failure analysis supported by multiple angles, full life cycle, and multiple information sources [6, 7]. In future, the prediction of pipeline failure model analyzes the failure factors of fragilities on big data, and uses Bayesian survival method by accident cause theory to predict life, and proposes that corresponding prediction models will be the hot pot of the research.

2. Characteristics and quantification of oil and gas pipeline failure

2.1 Characteristics of failure

The accident is an inevitable product of human industrial development and an important way in which the industry naturally chooses to eliminate it. It is also an important driving force for the industrial progress. In this chapter, leakage or certain economic losses of pipeline system as accidents or failure is defined. Many times accidents are described as failures and no distinction is made here. According to the pipeline accident statistics report [8–11], the causes of pipeline accidents are manifold, and most of them are caused by multiple factors. The industry has done a lot of analysis on the causes of pipeline accidents. However, in order to raise awareness of pipeline accidents, it is necessary to understand the accident intrinsically.

The accident has the following characteristics:

1. Inevitability: the accident will inevitably happen. It can only prevent accidents to the greatest extent, but it cannot prevent accidents from happening.

2. Irreversibility: any accident is irreversible, so it is more advantageous to increase the prediction and judgment of the accident than to ignore it.

3. Unknown: never know what will happen, unexpected accidents may occur at any time.

4. High probability: before the accident did not happen there will be lots similar but unhappen events. The types of accidents that have generally occurred will occur several times. The opposite is the small probability. For some type of small number incidents, it should be considered whether it is a low probability accident.

5. Not predictable: before the accident did not happen, it was considered that a very serious forecast may not be recognized. When an accident occurs, it cannot predict its consequences immediately. Dangerous factors may be safety, and unnoticed factors are often prone to accidents.

6. Recognize deviations: accidents, whether large or small, should be treated equally, but usually only serious accidents will be given attention. Experiments have found that people's ease of feeling about different risks is closely related to their emotional responses to these risks. Judgments and decisions can be influenced by emotions that happen without any awareness. The process of dealing with small accidents is feeling easy, after dealing with major accidents. The reverse order is feeling irritating, panicking, and easy to treat with mistakes. In fact, only focusing on major accidents is losing the chance of discovering problems from small accidents and avoiding major accidents. It is easy to occurrence of "probability neglect." There is a potential habit of coping with small risks, either completely ignored or over-emphasized and without buffers. "Availability bias" also affects the understanding of the accident. Such as public opinion, personal experiences, etc. will increase the impression of events, so that the brain is more likely to think of these things, so that these things are more likely to happen, but the reality may not be the case, this is only everyone's own understanding.

7. Tolerance is not uniform: the accident is unchanged, but the tolerance of the accident changes with time or place. When there is a need for war or survival, the tolerance for accidents rises; and when peace time and economic conditions are good, the tolerance for accidents decreases.

8. The simplicity and complexity of the cause: the cause for most of the accidents is very simple, but the process and factors that cause the accidents are very complicated.

9. Uncertainty of the cause. Many accident causes cannot be known. Perhaps, individual decisions are correct, but they do not determine that there will be no accidents.

10. The dichotomy of the accident: incidents include stages of occurrence and losses caused. For example, a pipeline leakage accident is divided into leakage and causes of loss. At the stage of occurrence, it includes leakage, diffusion, and explosion. The consequence phase is the serious loss caused by the formation of people, buildings, environments, etc. The first phase focuses on the direct cause of the leak, while the second phase analyzes the direct causes of serious consequences.

11. The ambiguity of accidents and affairs: an accident is an objectively occurring phenomenon that does not change; but as the situation changes, awareness and description of the accident will change accordingly. The level of awareness of accidents, response measures, and public opinion guidance determine the direction of affairs change..

12. The impetus of industrial progress: the accident will inevitably promote the improvement of management. Major accidents will become the turning point of cognition and management, and promote the improvement of technology and management level.

The above features are not all the features of the accident. With the deepening of the cognition of the accident, new features will continue to be recognized, and the existing features need to be further improved and supplemented. The characteristics of accidents are the essential attributes of accidents. However, this point is often overlooked, which leads to the deviation from the nature of accidents in analyzing and recognizing accidents, and results in erroneous analysis.

2.2 Accident probability and qualitative analysis

Whether or not all accidents should be investigated is the first problem in accident management decision-making. If this problem cannot be solved theoretically, it will result in unnecessary investigations and erroneous investigations, leading to erroneous management improvement measures.

If the occurrence of an accident is a low probability event or a random event, the causal explanation for it must be wrong. A small sample of the risk of error may be as high as 50%. In this case, if it is still handled according to the traditional high probability accident, it can easily cause greater losses [12]. The confirmation of the probability of an accident is the basis for deciding whether or not to conduct accident analysis and investigation. Therefore, the probability confirmation of the accident must be discussed before the accident investigation.

For random problems, the determination of the overall problem through a small number of accidents or random inspections is the key, which can be analyzed using the Larger Number Theorem. Bernoulli's theorem of large number is: let μ be the number of occurrences of event C in n independent experiments, and the probability of occurrence of event C in each test is P. Then, for any positive number ε, there is:

$$\lim_{n\to\infty}\left(\left|\frac{\mu_n}{n} - P\right| < \varepsilon\right) = 1 \qquad (1)$$

This theorem is a special case of Chebyshev's Law of Large Numbers, which means that when n is large enough, the frequency of occurrence of event C will be almost close to the probability of its occurrence, that is, the stability of the frequency. However, frequency is not equal to probability. As the theoretical basis for estimating the truth from statistics, the Large Number Theorem states that if the statistical sample is large enough, then the frequency of things can be infinitely close to its theoretical probability, that is, its "nature." It should be noted that "having a certain regularity" does not mean that a limited number of repeated trials will be taken, and the results obtained must have regularity. Even if a very large N is given, as long as $n < N$, it cannot be guaranteed. The frequency of occurrence of C is equal to the probability of its occurrence.

Generally, people misunderstand randomness as homogeneity: if something that has happened in the past has been uneven, it is mistakenly thought that things in the future will move toward "nearly uniform" direction. If an individual's misoperation like tossing coins is random, then the misoperation is not necessarily an individual subjective behavior. It may be due to the lack of certain standard constraints. It should be prudent to determine whether it is a personal problem, small probability problem, standard management problem, or ubiquitously large probability problem. For example, in the analysis of pipeline welding defects, the determination of whether a welder's welding problem or the welding quality of all pipeline construction management systems is a problem that needs to be clarified first.

If you decide to analyze the case, analyze it according to the law of small numbers. The law of small number is an important research result obtained by the Nobel Prize-winning behavior psychologists Amos Tversky and Daniel Kahneman in his study of "gambler's fallacy." It means that because of the limited human cognition of things, its behavior often shows heuristic biases and there is a problem of "spud," that is, the follow-up judgment result has a strong relationship with the starting point, and will be based on the observed several results to speculate on the later results, while ignoring the independence between each experiment.

The confirmation of the nature of the accident with small probabilities is the basis for deciding on the nature of the accident. If the survey is conducted without first determining whether it is a small probability event, there will be a significant impact of preemptive bias, leading to adverse consequences. Therefore, it should be judged first whether it is a small probability event. For a non-small probability event, a detailed accident investigation should be conducted and the improvement should be made based on the investigation results. In the investigation and analysis process, the investigation of the cause of the accident's occurrence and consequences, as well as the investigation of the cause of the accident, must also be based on the judgment of the nature of the accident with large probability or small probability.

The law of small numbers holds that if the sample is not large enough, then it may manifest itself in various extreme situations, and these conditions may not be related to the nature of the sample. You cannot rely solely on the judging person's own experience, and even add other people's experience to make judgments about things. The personal experience is limited. It is not possible to apply individual cases and large-scale statistics should be conducted. At the same time, before summing up the rules based on the sample results, try to avoid the conclusion of "anchoring" first, that is, you cannot first set a conclusion that is acceptable in mind and then look for a sample that supports the conclusion.

2.3 A quantitative description of the likelihood of an accident

Although there are many studies on the causes of accidents, there are few studies on how to quantify the possibility or status trend of accidents or failure. That is, there is little research on the quantitative assessment of the status of a system or object from the perspective of the probability of an accident when there is no accident or after an accident. Because risk is not measurable, the author tries to use the triad classifies, fragile, anti-fragile, and robust, proposed in the book anti-fragility, to quantify pipeline accidents or failures trend [13].

The concept of vulnerability is proposed by Keohane and Nye in "Power and Interdependence" [14]. It is mainly used to analyze the concept of international politics and assess the cost of changing the system of interdependence, that is, assess the severity of the consequences of violating or changing the rules. "Anti-fragile" believes that the triad classified of things is fragile, robust, and anti-fragile. Fragility is related to how a system suffers from the variability of its environment beyond a certain preset threshold, while anti-fragility refers to when it benefits from this variability—in a similar way to "vega" of an option or a nonlinear payoff, that is, its sensitivity to volatility or some similar measure of scale of a distribution [15]. That is, system becomes stronger or weaker after things such as fluctuations, pressure, etc., and other environmental or internal factors change. Specifically, fragile is a weak element in an object or system that may be exploited by the threat to cause damage. Once the fragility is successfully exploited, it may cause damage to the object or system. Fragile may exist in physical environment, organization, process, personnel, management, hardware, software and information, etc. Fragile refers to the trend of changes in things, such as volatility, randomness, pressure, etc. If an incident does not respond to volatility, randomness, and stress better, it means that the thing or system is vulnerable. A change trend that has no effect on the thing or system means that it is resilient or complete in response to fluctuations; if things exhibit greater flexibility in responding to fluctuations, an increase in benefits, or a decrease in damage, it means that the item is anti-fragile in response to fluctuations.

The fragile, anti-fragile, and toughness used to quantify the state of the accident system are only static physical states. The difference from traditional risk assessment is that the fragile is measurable, but the risk cannot be measured, and the measure of fragile is also different from risk prediction [13].

For oil and gas pipelines, in order to describe the trend of possible pipeline accidents, the entire system such as the physical state of the pipeline system and the management system can be described with the fragile, anti-fragile, and integrity. The choice of integrity instead of robust is based on the current state of the art of pipeline integrity management and assessment. Pipeline system fragile refers to the system that suffers from the fragility that may occur when pipeline internal and external factors change.

Fragility describes the possibility that a defect develops into an accident or an intact state develops into an accident. The anti-fragility of pipelines refers to changes in internal and external environmental factors, and even in the event of accidents, the pipeline's emergency measures, integrity management, etc. have been strengthened. Anti-fragility describes the development of pipeline systems in the direction of accident prevention. Integrity refers to the physical state of the pipeline and its ability to withstand changes in various factors. Fragility and anti-fragility are the status of the pipeline system, and its comprehensive indicators reflect the possibility of accidents in the pipeline system. Integrity means that the physical state of the pipeline is intact and its ability to withstand changes in various factors is strong. Vulnerability and anti-vulnerability are the status of the pipeline system, and its comprehensive indicators reflect the possibility of accidents in the pipeline system.

Entropy increase includes both the external factors that cause fragility and the internal factors in the system itself, which can reduce the integrity of the system. The entropy increase of the pipeline system makes the status of the pipeline system worse, and accidents are more likely to occur. The increase in entropy is controlled through management upgrades and maintenance inputs. If the entropy increase of the pipeline system fails to receive sufficient attention, it may increase the negative anti-fragile index, that is, the increase in entropy is a normal phenomenon, and the phenomenon of "warm boiled frog" occurs.

In the presence of known fragilities or hazards in the system, when external factors change, the factors that cause the accident are called trigger factors. For factors that are flawed, but do not cause an accident under conditions of external factors, are not triggers. The accident can happen because there are multiple triggers, or it can be the last trigger. The triggering factors may be related to time or maybe timeless, which is the direct cause of the accident. The trigger factor may be that a known fragility factor develops into an accident, or is an unoccupied or unrecognized fragility factor.

Therefore, the state of an accident that may occur in the pipeline can be described by the following formula:

$$A = F - AF \tag{2}$$

In the formula, A is the probability of accident; F is the probability of accident described by fragility; and $A\ F$ is anti-fragility.

Since it is the fragility of the triggering factor that determines whether an accident occurs or not, the fragility probability F_p described by the fragility is equal to the trigger fragility F_t, which is the direct cause of the accident. This results in:

$$F_p = F_t = max_i\{F_y, F_w\} \tag{3}$$

where F_y is the identified fragility factor and F_w is an unidentified fragility factor.

Because the fragility indicators of unknown factors cannot be identified and evaluated, the probability of an accident described by the fragility is generally equal to the maximum value of all known triggering vulnerabilities. F is a positive value. The larger the value is, the more fragile the physical state is and the more prone to accidents.

$$F = max_i\{F_y\} \tag{4}$$

In the above model description, there will be two cases:

1. One of the known factors develops to an accident.

For example, assuming that a pipe has 10 conditions that can lead to the occurrence of cracks in the girth weld, the current girth weld has 9 known factors, and the existence of these 9 factors does not cause an accident. When one of these 9 factors changes, it causes an accident. This factor may be an internal or external factor. It may be that the weld crack defect grows to the extent that it cannot carry the normal operating pressure of the pipeline, or it may be that the geological displacement causes an increase in the external bending stress.

2. Unknown factors triggers to accident.

For example, among all the factors of the cause of the accident, the known factors of vulnerability did not change further, but the occurrence of unknown or missing triggers led to the accident. Supposing that a pipeline with 10 factors can cause cracks in the girth weld of a pipeline, the current girth weld has 9 factors condition, and if the tenth factor condition is met, such as the internal pressure rises due to misoperation as a trigger, an accident occurs.

Therefore, the probability of an accident is determined by the fragility factor that has the highest probability. When conducting risk management, you should focus on the most probable fragility factors, rather than assessing and managing every risk of the entire system.

When the triggering factor is timeless, traditional risk assessment may think that its risk is high, but its qualitative change from high-risk to accident may never happen, or it may happen at any time. Therefore, the probability that the triggering factor break determines the risk of the entire system, rather than judging and counting the risk based on experience, and more importantly, the fragility status determines the risk.

When the probability of fragility factors cannot be identified technically, it can be assumed that there are many existing fragility factors in the system. Once the conditions change, the probability of accident of both types is high. The "accounting rate" of factors corresponds to the probability of an accident.

Therefore, according to the accidental "tree-type" theory [5], it is assumed that the largest factor causing a certain type of pipeline accident is m, and the state is found to have k fragile factors. According to the occupation rate, the probability of occurrence of the accident is k/m.

Therefore, for a multi-factor i that may result in fragility, the fragility can be quantified using the following formula:

$$F = \Pi_i\left\{\frac{k}{m}\right\} \tag{5}$$

In the formula, *i* is the number of risk factors that cause fragility, for pipeline that will be including the total number of corrosion, cracks, third-party damage, etc.

Assume that there may be 10 factors that may cause cracking of girth welds in a pipeline, including the lake of fusion, crack, repair welding, dead ends' welding, etc. Through the risk identification, it was found that the weld has seven fragile factors such as crack, but the assessment results indicate that the pipeline can continue to be serviced under the existing pressure and other conditions. The triggering factor may be the growth of crack, or it may be the eighth or ninth newly added fragility, such as the external load caused by geological displacement, overpressure by misoperation. Under fragility conditions that cannot be clearly determined, it is believed that the higher the F value, the easier the break. When internal and external environments change, accidents are more likely to occur. Comprehensive calculation of F-values and consequences can give the risk of the pipeline system.

According to the definition of anti-fragility:

$$A_f = M_g - L_f \tag{6}$$

In the formula, M_g is the management improvement effect and L_f is the negative effect.

When the effect of management improvement is greater than the negative effect, A_f is positive; otherwise, it is negative.

For oil and gas pipelines, its integrity is defined as:

$$I = 1 - F \tag{7}$$

where I is the integrity of the pipeline, that is, the physical state of the pipeline that meets the design and operational requirements.

Fragility describes the state of the pipeline system, which is only the physical state at a specific point in time. When the factor conditions change, the fragility will be changed. When conditions change independently of time, the fragility is also independent of time. Therefore, the fragility cannot be used to predict the probability of an accident. The likelihood of accidents is the degree of difficulty or likelihood of triggering. So the concept of fragility is different from that of the traditional pipeline accident.

In general, the ratio calculated from accident statistics is used as the probability of an accident, which is a calculation or statistic. Muhlbauer, in his newly published book in 2015, Pipeline Risk Assessment—The Definitive Approach and Its Role in Risk Management, believes that the most useful definition of probability is the level of trust [16], including engineering judgments, expert opinions, and potential risks that have been evaluated understanding of physical phenomena.

The traditional risk assessment is to use the experience and statistics of past accidents to predict the possibility of accidents occurring in the future, rather than describe the physical state of the system. This type of risk assessment method often fails to answer the question: "High-risk pipelines have not been involved in high-risk accidents for a long time," which has seriously affected the degree of recognition and acceptance of risks by some public, engineers, and senior management personnel.

If we assess and describe the state of the system based on the fragility, we can more scientifically describe the system's possible accidents, and more clearly reflect the process and possibility of the development of high-accident factors as accidents. If on this basis, combined with the consequences of accidents to comprehensively measure, it can reflect the state risk of the pipeline system, and the conditions and factors that lead to the establishment of an accident or risk will be clearer.

Fragility is not reliability. Reliability is generally an indicator for non-repairable systems. It is the ability to continuously provide normal use functions through parallel systems in the absence of faults or under fault conditions.

According to Mohitpour et al.'s Pipeline Operation and Maintenance: A Practical Approach, the reliability of the pipe system is considered to be the degree of reliability of the media to be safely transported through pipes within a given time [17]. Some pipeline operators calculate the reliability of a pipeline by using the total amount of time that a pipeline can deliver a certain amount of medium without failure during a certain period of time. It can be seen that the reliability assessment is an assessment of the integrated system such as process and system status, and the fragility is only a key factor of reliability.

3. "Tree-type" model of accident cause theory

In areas such as the oil and gas industry and the construction industry, accidents such as leaks and explosions of oil and gas pipelines, and the collapse of buildings and bridges often occur with serious consequences. In the face of accidents, people are always thinking about the causes of accidents. This has led to the theory of accident causes and has become the theoretical basis for understanding accidents [18].

The accident-causing theory promoted the recognition of the objective world and reduced the harm of accidents to humans, which is a direct reflection of human technological progress. The occurrence of an accident, especially the occurrence of a casualty accident, is random and accidental. It is caused by many factors, and there are intricate and complex relationships among various factors. The theory of accident cause is to find out the inevitability of the accident from the random and accidental accident, understand the law of the accident, and eliminate the accident in the bud.

At present, the more-recognized causality theory focuses on the analysis of the unstable conditions under the conditions of accidents, ignoring its historical background and causing a great deal of information loss. For this reason, based on the analysis of numerous accidents, a new theory of accident causation was proposed, and the core content and new criteria for analyzing the causes of accidents were systematically presented. Then, an accident failure analysis method was proposed based on this theory include analysis process and logic analysis model in order to be able to improve the drawbacks of the traditional theory, improve the cognition of the cause of the accident.

Due to the limitation of professional technology, the "tree-type" accident cause theory proposed below is mainly aimed at oil and gas pipelines and building structure accidents, but it also has reference value for the analysis of accident causes in other area.

3.1 Traditional accident cause theory

Accident cause theory is the product of the development of productive forces to a certain stage, especially the change of production methods and the change of people's status in the production process, causing changes in people's safety concepts, resulting in different accident cause theory. With the continuous improvement of the level of production technology and the deepening awareness of accidents, the accident cause theory has also been continuously improved and developed, and has gradually become an important theoretical basis for understanding the accident process and preventing accidents [19]. To sum up, it mainly experienced three historical periods of single factor cause theory, two-factor cause theory and system cause theory.

The accidental cause theory is gradually developed in the process of modern industrial development. In its development process, Adams' accident causal chain theory, trajectory intersection theory, human factor system theory and synthesis are the comprehensive theory. The difference in the core content of these theories reflects the continuous improvement of the depth of awareness of accidents. The theoretical analysis of comprehensive causes is the most scientific, and it comprehensively covers other theoretical contents.

The analysis of several current classical theory of cause of accidents has certain limitations and deficiencies. The core of the theory is mainly about insecure physical state of things and the unsafe behavior of people. Failed to grasp the true features and root causes of the accident. The effects of incentives on accidents were not taken into consideration, and the comprehensive role of multiple factors was not considered as the key to the accident. What is more important is that the comprehensive relationship between accidents and all aspects of the entire life cycle and the environment is not taken into account, and the possibility of technical ambiguity is not taken into consideration. Therefore, the problem cannot be fundamentally explained.

The traditional accident cause theory believes that an accident is caused by the cross between the unsafe state of the object and the unsafe behavior of the person. Obviously, it describes only an idealized simple concept. This theory cannot explain the situation in which the two are crossed without accidents. At the same time, it cannot explain the situation where the two do not exist or do not cross, but the accident still occurs. For this reason, the cause of the accident cannot be really identified only by the crossing of the trajectory.

The use of traditional theories to analyze accidents can easily cause certain causes of omissions in many aspects. In particular, the underlying causes of analysis are prone to large deviations. For example, in the building structure and pipeline industry, the occurrence of accidents is affected by many factors. It is necessary to consider all aspects of the life cycle such as planning, design, construction, and operation. Often, the unsafe physical state of things and people's unsafe behavior are not fundamental to the accident. The traditional cause theories, therefore, are not recommended as the core of the accident causal analysis.

3.2 "Tree-type" model of accident cause theory

Before discussing the causal theory of the "tree-type" pipeline accident proposed in this paper, several basic definitions are discussed.

3.2.1 Definition

Root cause: it refers to the root cause of things or the most essential reason that causes things to change. It is the most important reason that plays a key role and determines the role of many things that cause things to change.

Indirect cause: it is a factor that cannot play a leading role. It only plays a role in the transfer.

The direct cause (usually also the inducement): it refers to the most direct promotion of the occurrence of things and the direct contact with the occurrence of accidents. It is the near-term surface expression that causes changes in the development of things. It does not go through intermediary or intermediary links. Generally, it is to analyze the closest factor in time relationship or logical relationship.

3.2.2 Fundamental arguments for the cause of the "tree-type" accident

The principle of inevitability: the occurrence of an accident is inevitable; there is no accident that occurs as contingency; only the probability is different. With the increase in awareness of the causes of accidents, it can be seen that when certain conditions are met, accidents are inevitable and can be predicted and prevented.

Direct consequence: the direct consequence of an accident is structural failure, leakage or Destroyed of device etc. The failure includes failure of ontology function, failure of auxiliary facilities, etc., while the loss refers to the state of physical corrosion loss, structural protection loss, and other deviations from the original design requirements.

Indirect consequences: the indirect consequences of an accident are casualties, environmental pollution, or property damage.

Direct cause: if accidents occur due to inadequate or unfavorable changes in the natural environment, management (human), and nature (process, corrosion, and fatigue), this change can be considered as a direct and direct cause of the accident. Any factor may break the balance. There is no concept of quantity.

Indirect reasons: the insecure status of objects and human unsafe behaviors, and the factors that create the necessary conditions for accidents are indirect causes. The insecurity of things and the unsafe behavior of people should all be attributed to the management elements. It is just a direct manifestation of various kinds of missing in time coordinates, rather than the direct cause of the accident. It is an indirect cause. The intersection or non-intersection of the two may cause accidents, and the intersection between the two may not cause accidents. Whether or not the accident is affected by the amount of crossover, incentives, etc. Although accidents occur when the insecurity of things and people's unsafe behaviors intersect, it does not mean that they are the root cause of the accident. The lack of deficiencies in each link of the entire life cycle or the inadequacy of reliability over time may add up to a certain extent, which may lead to accidents. If the absence of the latter link leads to the lack of the previous link, it may have a continuation effect. If the latter link corrects the lack of the previous link in a timely manner, the accident may be controlled in this link. There is no lack of a link itself cannot make up for the lack of upstream links, can only reduce the contribution rate to the accident. Designs or requirements based on a large number of preconditions and designs without loss of control are missing.

Quantification of indirect causes: the lack of various links (insecure state of matter, lack of management) varies according to external factors (administrative insecurity, natural environment, process change, etc.) under different conditions, and needs to be based on external conditions. Determine whether the absence will lead to an accident. Whether it is a single link or multiple links is the key to causing an accident. It is necessary to analyze the degree of lack of each single link, and the degree of influence of comprehensive factors and external incentives needs to be considered. The unsafe state of things and the unsafe behavior of human beings cannot be cut off, and it is impossible to eliminate them completely. It is also possible that the factors are considered to be in a state of safety and the accidents are caused by superposition.

Root causes: it refers to the nature and root causes of various factors. It is the system, mechanism, law, humanities, education, economy, standards, norms, management methods, assessment mechanism, and technical cognitive level that lead to the lack of the main body, lack of management, etc. The phenomenon is generated, which accumulates and develops into an accident. This is the root cause.

Quantification of root causes: different social factors, life cycle factors, and direct influence factors are the key to the possibility of an accident.

The unique characteristics of accidents: the types of accidents are different, and the number, distribution, severity, and external incentives of the links are different. It is necessary to perform logical analysis according to the actual situation.

3.2.3 Logical model of the cause of "tree-type" accidents

The "tree-type" accident causation theory is based on the conditions of big data [6], combined with case analysis and verification, and proposes a new theory of accident causation. Its purpose is to grasp some potential laws of accidents are followed. The accidents that have occurred are tracked. The root causes and evolution of the accidents are studied and analyzed to understand the tendency and inherent laws of the accidents so as to find the potential or possibility of accidents in the future. Therefore, adopting corresponding measures to prevent or eliminate accidents and avoid "reproducing" accidents is a philosophy of accident-free [2].

According to the above arguments, the causes of accidents can be simulated using the logic of "tree-type" like the life of tree: accidents are just like a dead tree, broken branches, and other processes. When standards, norms, education, mechanisms, and other fundamental issues are similar to bad soil environmental impacts, the health of the tree determines its basic nature. Design, construction, and other aspects will affect the strength and ability of the entire tree. The various production processes of operations are similar to the leaves of trees, and play a role and contribution to the whole in the sunlight of management. If problems arise in management, the entire process will be affected, causing the trees to wither or be blown off in the breeze. Process changes and changes in the elements of the natural environment are the external environments in which trees live. Just as seasonal changes and wind and rain are the external influences on the health of trees, they are the external environments that a healthy body should bear. The changes are the internal defects that lead to the development of accidents. It is the direct cause. Based on the above, the "tree-type" simulation was conducted for the pipeline industry as **Figure 1**.

3.2.4 Application analysis process of "tree birth" theory

The value mining of accidents is only the tip of the iceberg. A large amount of regular and valuable information has not been fully analyzed and identified, resulting in the loss of precious wealth. Under the conditions of big data, the awareness of accidents has changed and can be fully realized. Digging out the "long-tailed" factors that may cause accidents can also fully avoid the "survivors' rule" [20].

The "tree-type" pipeline accident causation theory aims to reveal the relationship and laws between accidents and life cycle influencing factors, so as to feed back to each stage of the life cycle and prevent the repetition of similar events. According to the principle proposed by the basic arguments of the "tree-type" accident cause theory, the causal analysis was carried out in order as **Table 1**.

The accident causation theory is the theoretical basis of the accident investigation When the accident investigation is conducted, data collection and analysis are performed according to the various factors and logic involved in the theory. So the pipeline big data era lets the full life cycle of "tree-type" theory to be operated. Under big data conditions, how to recognize pipeline accidents and risks gives a different point of view than traditional. The change of risk assessment under big data conditions is inevitable. The prediction of pipe girth accident model based on big data and "tree-type" theory is only an application example, and its conclusion has shown that this analysis method gives more reasonable conclusions [2]. It is easier to identify the cause of the accident and give suggestions for management improvement.

This theory is also the theoretical basis for risk identification and assessment. Under the conditions of big data, by analyzing the key factors leading to accidents, and calculating the ratios of potential factors and causes, the possibility of accidents can be calculated and the key factor of the mitigation can be given. Based on this idea, a new and systematic risk assessment model can be constructed.

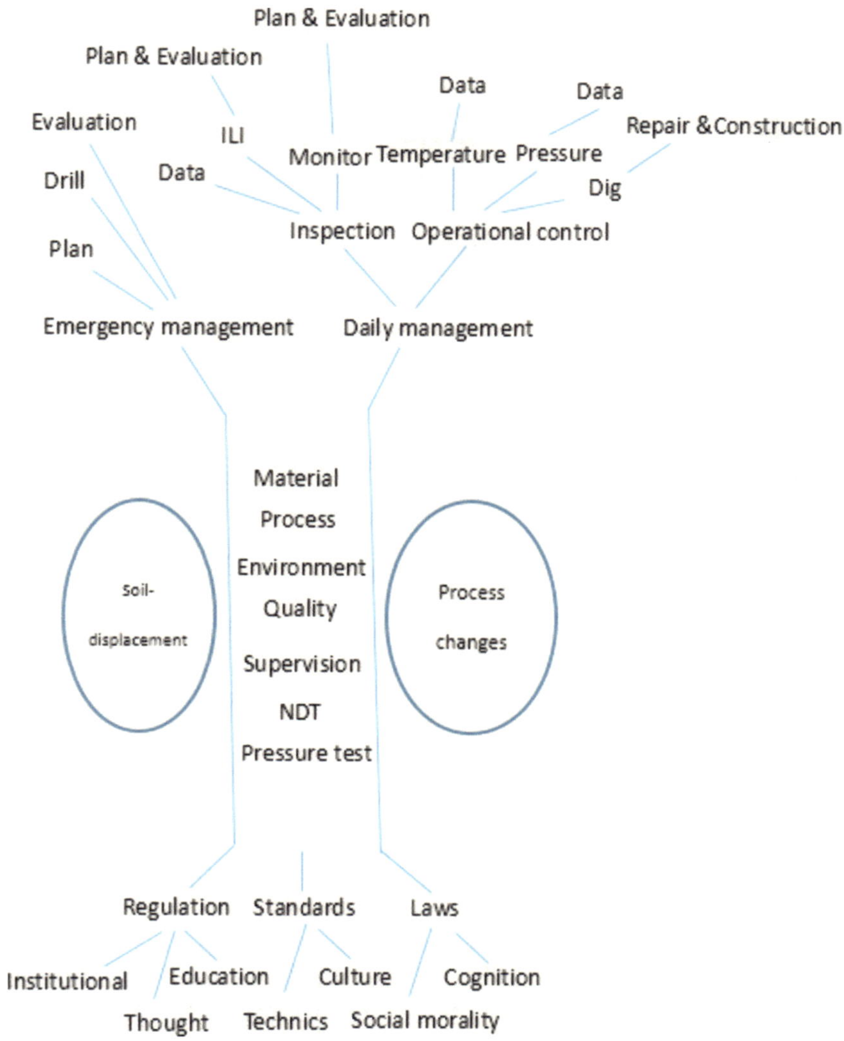

Figure 1.
Tree-type simulation of pipeline accident cause theory.

3.3 Casing study

3.3.1 San Bruno pipeline rupture

On September, 2010, a gas pipeline operated by Pacific Gas and Electric Company (PG&E) ruptured under the suburb of San Bruno, California, USA. The rupture produced a crater about 72 feet long by 26 feet wide. The released natural gas ignited, resulting in a fire that destroyed 38 homes and damaged 70. Eight people were killed, many were injured, and many more were evacuated from the area [21]. Applying the tree-type theory to analyze the accident can more systematically identify the cause of the accident as **Table 2**.

3.3.2 Pipeline pigging stopping accident

A pigging stop occurred during a pigging operation for internal inspection of a pipeline: the first pig was parked at the elbow. After the analysis was performed using traditional analysis methods, specific improvement measures were

No.	Steps	Course analysis
1	The principle of inevitability	Define the inevitability principle of accident analysis and confirm that it is a small probability event or a single event.
2	Direct consequence	Investigation to clarify the direct performance of the accident.
3	Indirect consequences	Investigate indirect performance and losses, such as casualties, property damage, etc.
4	Direct cause	Analyze the direct cause and find out the factors that break the balance. There may be multiple factors.
5	Indirect cause	Identify unsafe status of objects and unsafe behavior of people, analyze cross history and quantity, find data of full life cycle, analyze possible lacks of each link, and find indirect causes.
6	Quantification of indirect causes	Quantify the contribution of indirect causes to accidents.
7	Root causes	The various factors of the whole life cycle link to find out the root cause.
8	Quantification of root causes	Quantify different social factors, life cycle factors, and direct influence factors.
9	The unique characteristics of accidents	Analyze the uniqueness of various types of accidents according to the differences among various factors of different accidents.

Table 1.
Tree-type theory analysis process.

implemented, but after restarting the operation, a second card stop accident occurred and the pigs stopped at the "t" section. According to the traditional accident analysis method, the root cause of the first pig stopping accident was analyzed. The pig cup was cracked by a 1.5 D (D-pipe diameter) elbow, causing the pig to lose its power and stop. The existing management problem is that there is no careful analysis of pipeline data, and the information obtained does not match the facts. According to the results of this analysis, the operator shortened the pig to enable the pig to pass through the low-rate elbow. However, due to the shortened length, the pig was passing through the tee because the length of the three-way flower hole was greater than the length of the pig. The tube device drives the distance between the cups before and after, losing power, and stopping. According to the traditional accident analysis method, it is believed that the current stoppage is due to the pigging device being too short. The reason for management is still that the pipe data is not carefully analyzed. The information obtained is inconsistent with the facts, and the management reason cannot be eliminated because of the history of the pipeline construction. As a result, the technology and management are not standardized during construction, and no comprehensive and accurate pipeline information can be obtained. If we continue to follow the traditional methods of accident analysis and treat similar problems, not only can we not really avoid the occurrence of accidents, but we also think that there are problems in management that have a greater impact on managers. Therefore, we should not simply think that there are problems with management.

According to the "tree-type" causal analysis method, the occurrence of a branch similar to the breakage of a branch, the common cause of multiple breakage of different causes is the irregular growth of branches, self-elimination in the natural environment, which is basically the root soil, trunk, etc. There are hidden dangers caused by the growth of irregular branches.

The direct consequence is that the pigs are stuck in the pipeline, causing minor production impacts. Increasing the pig pushing operation is an indirect

No.	Steps	Course analysis	Tree-type
1	The principle of inevitability	Define the inevitability principle of accident analysis and confirm that it is a small probability event or a single event. The dichotomy of an accident should be considered when analyzing the causes and consequences.	
2	Direct consequence	The rupture of the pipeline.	Dead or broken of the tree.
3	Indirect consequences	Casualties and property losses.	Broken tree brings other losses.
4	Direct cause	A pressure increase stemming from poorly planned electrical work at the Milpitas Terminal.	Decayed tree trunk breaks under strong wind.
5	Indirect cause	Inadequate quality assurance and quality control in 1956 during its Line 132 relocation project, which allowed the installation of a substandard and poorly welded pipe section with a visible seam weld flaw that, over time grew to a critical size, causing the pipeline to rupture.	The soil on which the tree grows and the growth environment is poor, results in the lack of robustness of the tree.
6	Quantification of indirect causes	If you quantify the contribution of each link to the accident, construction may be the most important factor.	Just as the tree has decayed in its trunk during its growth, the decayed trunk has the largest contribution to the breakage of the tree.
7	Root causes	It was inextricably linked with the social environment and technology limited at that time. Despite the fact that education, culture, and technology have been greatly improved over the past, there is no good technology of integrity management performance measurement lead weld defect grow to rupture.	The roots of the tree grow poorly in soil (social environment, culture, ideas, etc.) and lack of sunlight (laws, standards, regulations, etc.), causing them to grow into unhealthy and irregular trees such as rot. When the external environment changes, it is broken. The cause of the rot of its trunk is the root cause of the broken tree.
8	Quantification of root causes	In the pipeline's life cycle, neglected construction quality, poor of integrity management plan and inspection during operations, and weak safety awareness are the reasons that inevitably lead to accidents and consequences are extremely serious. Quantify the contribution of each link to accidents, and the lack of safety awareness and supervision is the most important factor.	Quantitative analysis of the contribution of various aspects of the decay of the trunk, including soil problems, lack of sunlight and other factors.
9	The unique characteristics of accidents	There is only one San Bruno pipeline rupture accident in the world, but similar problems exist in other pipelines. The accident has its own unique cause, but it is also universal. Its occurrence is inevitable.	

Table 2.
Tree-type theory analysis process of San Bruno pipeline rupture.

consequence, similar to the breaking of the branches; the incentive is for the pigging effect, and the investment size is larger during the pigging process. The pig-cleaner exceeds the requirements that the pipeline can pass, just as the decaying branch naturally breaks due to an increase in its own weight when it grows up; the pipeline is in a state of such objects from construction to operation for a long time, but it cannot be simply defined as unsafe status.

The indirect reason for the first stop is that pigs are designed according to the passing capacity of 4 D, but the actual pipe elbows reach a minimum of 1.5 D. In the absence of a determination of the root cause, only the passing capacity of the pigs is improved. The length of the pig was shortened to enable it to pass through a 1.5 D elbow, but it stopped at the tee section again during the pigging process. The cause of this short pig stoppage was too low a ballast flow. The indirect reason was short of pig and the tee flower hole is too long, causing the pig to lose its power.

There are certain management problems in the two stops. Without the tee size data, there is no systematic assessment for it to be passable. However, the root cause of the management problem is that the pipeline data it is based on is inconsistent with the facts. It is a pipeline construction background such as the social environment, technical level, and standards that cause the irregular pipeline were the root causes of the stop.

The above case also show that the presence of elbows and tee features in the pipeline cannot simply be considered as an unsafe state of things, nor can pigging operations is regarded as unsafe acts of people, but still a stuck pig accident occurred. Through the use of the "tree-type" theory, a systematic analysis of the factors that caused the pig stop and find out the root causes can recognize that the two pig stop accidents are caused by historical reasons, and there are deviations in the pipeline data on which management is based. Only by implementing pigging operations in accordance with the most stringent pipeline data, gradually reassessing pipeline data, and making decisions based on real data, can effectively avoid the pig stoppage.

At the same time, in the principle of "tree-type" accidents, the understanding of "insecurity of things and unsafe behavior of people" is deepened, that is, both are variables that lead to accidents. When they appear crosswise, they may happen and also may not happen. Only by analyzing the historical background and conditions of the pipeline's manufacturing, construction, etc., factors can the root cause of the accident to be identified. If the traditional accident analysis method is used, the root cause cannot be systematically and comprehensively analyzed, which may lead to the repetition of similar accidents.

The application of "tree-type" accident cause theory will raise the level of risk analysis and response at a higher level to mitigate risk and prevent accident. It is the key to the integrity management of oil and gas pipelines and building structures and technological progress, as well as the theoretical basis for achieving "prevention" [22]. At the same time, this theory systematically gives the core content of the analysis of the accidents causes. It can be widely applied to the cognition of industrial accidents such as building structures and oil and gas pressure pipelines. It can also build risk assessment methods based on this theory and is a risk assessment technology in the future as a new research way [23].

4. Summary and conclusions

The occurrence of non-leakage accidents is the goal of pipeline operation. Research on accidents cause theory is the theoretical basis for accident prevention. A clear understanding of the nature of accidents is the basis for the analysis and management of accidents and it will help to improve the technical level of managing oil and gas pipelines. The author's research raises the following awareness to help prevent pipeline accidents:

1. The essential feature of an accident is the clear recognition of the basis of the accident. In the analysis of accidents, inevitability, deviation, and dichotomy are all characteristics that are easily overlooked or forgotten. Accidents and events are often confused. This article lists only part of the characteristics of the accident. There will be other features that will be recognized, observed, and accepted gradually in the future.

2. After an accident, whether it is a low probability or high probability accident, or the nature of a random accident, is the basis for determining whether the accident needs to be investigated and analyzed. Cognition of accident samples is the basis for improving accident awareness. The characterization of the accident as a small probability event or a universal problem is the basis for making accident investigations and accountability. Accurate characterization can avoid incorrect conclusions due to incorrect investigation and analysis, and affect management improvement.

3. The quantitative description of an accident is different from the current traditional method of recording accident descriptions, and it is the theoretical basis for improving accident analysis and cognition. The concepts such as fragility are used to quantify the accident probability, and the differences from traditional risk assessment techniques are analyzed. This not only improves the level of awareness of accidents, but also has a certain guiding role in improving the direction of risk assessment techniques.

4. The author proposes to describe the possibility of system accidents with fragility, anti-fragility and integrity. This is the first time that the logical relationship and mathematical description of the accident analysis have been given, and it is believed that the fragility description is superior to the traditional risk assessment and can better reflect the system status and identify the key factors leading to the accident.

5. Traditional accident cause theory makes people pay more attention to the insecure state of the object and the unsafe behavior of the person, and treats the directly presented problem as a root cause. The "tree-type" accident cause theory gives a systematic analysis of the logic, ideas and methods of accident analysis. It is a comprehensive analysis of various factors, which helps to find the causes of accidents, and distinguishes between the direct causes and indirect causes are an effective supplement to the theory of accident causes.

Using the "tree-type" accident causality theory analysis method, it is easier to understand the logical relationship and process of the analysis of the causality of accidents. Based on the principles and processes proposed by the "tree-type" accident causation theory, the use of simulated logical relationships can be clearly understood the causes of accidents and avoid take indirect causes as root causes. The factors that lead to the accident, the logical relationship, and the contribution to the tree breakage that are analyzed using this method can fully identify the risk of the accident and how to control the risk, so as to effectively prevent it.

This chapter discusses that the cause of accident is an older and traditional research topic. The causes of accident and description methods of features, etc., are not the hot topics in recent years. However, the cause of the accident is the theoretical basis for pipeline integrity management. As the author discusses in the book "Research and Practice on Pipeline Integrity Management in the Eastern Pipeline Network" [3], the core of pipeline integrity management is accident prevention, so what is the accident, what causes accidents, and why accidents and other issues are the core of research and promotion of pipeline integrity technology. Only by clearly knowing the causes of accidents can we identify potential risks and assess their vulnerability. Therefore, the study of the theory of accidental causes is the theoretical basis for improving the integrity of pipelines. It is recommended that in-depth study be conducted. The consequences of pipeline accidents are great harm to environments and safety. With the large-scale application of high-pressure and high-grade steels pipe, scientific accident cause theories are needed to support and identify the causes of various accidents under new material and new process conditions. Protect and assessments the fragile of pipeline to address urgent pipeline operation needs.

In the future, it is necessary to continue to study the characteristics of accidents and improve the theory of accident causes. It will be a longer-term task to study how to use system fragility assessment instead of risk assessment based on big data. At present, the recognition of accident features, the causation theory, and the vulnerability assessment system based on big data are still incomplete, and its need to be improved. However, with the accumulation of big data, new methods such as quantitative accidents based on the use of fragility are available. These new technologies will be recognized and developed.

Author details

Qingshan Feng
PetroChina Pipeline Company, Langfang, China

*Address all correspondence to: qsfeng@petrochina.com.cn

References

[1] Feng Q, Dai L. Revelation of Oil and Gas Pipeline Failure. Beijing: China Building Material Press; 2016. ISBN: 9787516013045

[2] Feng Q. New concept based on big data to pipeline risk management. Pipeline and Gas Journal. 2015;**242**(3):P65-67

[3] Feng Q. Research and Practice on Pipeline Integrity Management in the Eastern Pipeline Network. Beijing: Petroleum Industry Press; 2016. ISBN: 9787518312108

[4] Feng Q. Theoretical study on characteristics and quantification of oil and gas pipeline failure. Oil & Gas Storage and Transportation. 2017;**36**(4):369-374

[5] Feng Q. "Tree type" model of accident cause theory. Oil & Gas Storage and Transportation. 2014;**33**(2):115-120 128

[6] Mayer-Schonberger V, Cukier K. Big Data: A Revolution That Will Transform How We Live, Work and Think. Boston New York, US: Houghton Mifflin Harcourt Publishing Company; 2013

[7] Feng Q. A thought over the pipeline risk assessment methods based on Big Data. Oil & Gas Storage and Transportation. 2014;**33**(5):457-461

[8] Pipeline and Hazardous Materials Safety Administration. All Reported Pipeline Incidents. 2017. Available from: http://www.phmsa.dot.gov/pipeline/library/data-stats [Accessed: February 16, 2017]

[9] National Transportation Safety Board. Pipeline Accident Reports. 2017. Available from: https://www.ntsb.gov/investigations/accidentreports/pages/pipeline.aspx [Accessed: February 16, 2017]

[10] Van Denbrand D. European Gas Pipeline Incident Data Group Reports. 2017. Available from: https://www.egig.eu/reports [Accessed: February 16, 2017]

[11] National Energy Board. Pipeline Ruptures. 2017. Available from: https://www.neb-one.gc.ca/sftnvrnmnt/sft/pplnrptr/index-eng.html [Accessed: February 16, 2017]

[12] Kahneman D. Thinking, Fast and Slow. New York: Farrar, Straus and Giroux; 2011. ISBN: 978-0374275631

[13] Taleb NN. Anti-Fragility. Longman: Random House; 2000. ISBN: 1-400-06782-0

[14] Keohane OR, Nye JS. Power and Interdependence. New York. ISBN: 978-0205082919

[15] Taleb NN, Douady R. Mathematical definition, mapping, and detection of (anti)fragility. Journal of Quantitative Finance. 2013;**13**(11):1677-1689

[16] Muhlbauer WK. Pipeline Risk Assessment: The Definitive Approach and Its Role in Risk Management. Houston: Expert Publishing, LLC; 2015. pp. 18-24

[17] Mohitpour M, Szabo J, Hardeveld TV. A Practical Approach for Pipeline Operation & Maintenance. New York: ASME; 2012. ISBN: 9780791859605

[18] Chen X, Feng Q, Shui B, Li B, Hao J, Teng T. The development and challenges of pipeline failure analysis in China. In: 7th International Pipeline Conference (IPC); 2008. DOI: 10.1115/IPC2008-64503

[19] Ibn Awal Z, Hasegawa K. A study on accident theories and application to maritime accident. Procedia Engineering. 2017;**194**:298-306

[20] John T. Pipeline Pigging and Integrity Technology. New York: Clarion Technical Publishers; 2003

[21] National Transportation Safety Board report NTSB/PAR-11/01, Pipeline Accident Report, Pacific Gas and Electric Company, Natural Gas Transmission Pipeline Rupture and Fire, San Bruno, California. September 9, 2010. Available from: https://www.ntsb.gov/investigations/AccidentReports/

[22] Feng Q. Three-axis high-resolution MFL internal inspection technology for in-service pipeline. Oil and Gas Storage and Transportation. 2009;**28**(10):72-75

[23] Cao Y, Xu W. System engineering risk assessment method research progress. Engineering and Science. 2005;7(6):88-95

Chapter 3

Stress Corrosion Cracking Damages

Alireza Khalifeh

Abstract

Stress corrosion cracking (SCC) is the formation and growth of crack through materials subjected to tensile stress and a specific corrosive medium. It can lead to unexpected sudden failure of normally ductile metals. Metal-environment combinations susceptible to cracking are specific. This means that all environments do not cause SCC on all of the alloys. Additionally, the environments that cause this kind of cracking have little corrosion effect on the alloy in normal conditions. In certain states, unwanted environmental and metallurgical changes have occurred and provide the metal-environment combination sensitive to SCC. The SCC sites on the metal surfaces may not be visible by visual inspection, while metal parts are being filled with microscopic cracks. These invisible cracks progress rapidly and lead the component and structures to catastrophic failures. In this chapter, the incidence of SCC on important industrial alloys from the chemical, metallurgical, and mechanical point of view is discussed.

Keywords: stress corrosion cracking, environments, materials, mechanisms, fracture mechanic

1. Introduction

Stress corrosion cracking (SCC) in chemical, petrochemical, and power plant industries is an insidious form of corrosion, which causes a lot of financial losses and human damages [1–5]. This phenomenon is associated with a combination of tensile stress, environment, and some metallurgical conditions as described in **Figure 1**.

During stress corrosion cracking, the metal or alloy is virtually unattacked over most of its surface, while fine and branch cracks progress through the bulk of material [6]. It is shown in **Figure 2**. This cracking phenomenon has serious consequences since it can occur under stresses much lower than design stresses and lead the equipment and structures to premature failures [7–11].

Stress corrosion cracking starts from corrosion sites at the material surfaces and progresses into a brittle manner. The process of cracking is not strictly a mechanical process, as the corrosivity of the environment strongly affects the fracture mode. Both intergranular and transgranular stress corrosion cracking are observed. Intergranular cracking proceeds along grain boundaries, while transgranular cracking advances without apparent preference for boundaries [12]. An example of stress corrosion cracking in which the crack has progressed in both intergranular and transgranular paths is shown in **Figure 3**. The development mode of cracking depends on the composition and microstructure of the material and environment.

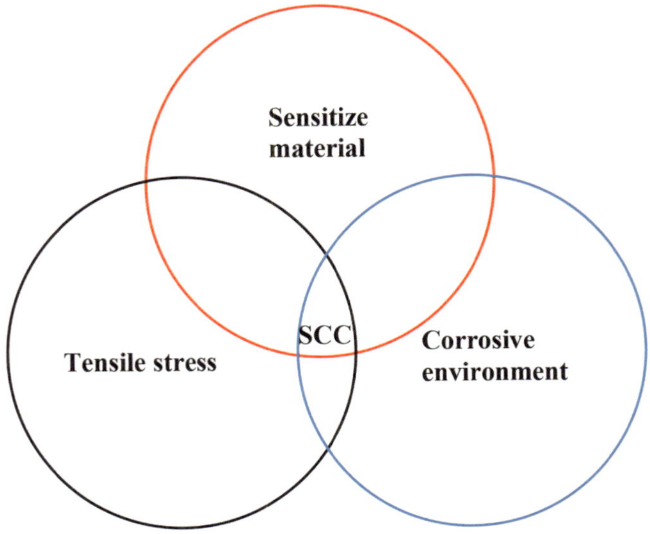

Figure 1.
The essential requirements for SCC.

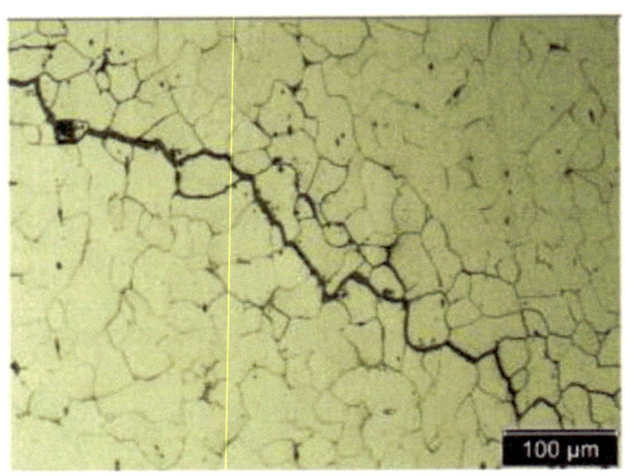

Figure 2.
Crack development in carbon steel exposed to nitrate solution.

In this chapter, the conditions for the occurrence of SCC are first introduced. Then, the stress corrosion cracking mechanism for various materials in conditions that are susceptible is discussed in detail. The design of industrial structures and components is usually based on tensile properties, which have many disadvantages. So, the science of fracture mechanics applies in the situations prone to SCC because of the inevitability of manufacturing and service defects in materials and for considering the role of such imperfections. Methods of prevention based on corrosion science and empirical data are presented. Finally, practical examples are given to better understand the issue.

2. Requirements for SCC

Not all metal-environment combinations are susceptible to cracking. In other words, the environment for occurrences of SCC for each metal or alloys is specific. Also, the resources of stress for each case of failure may be different.

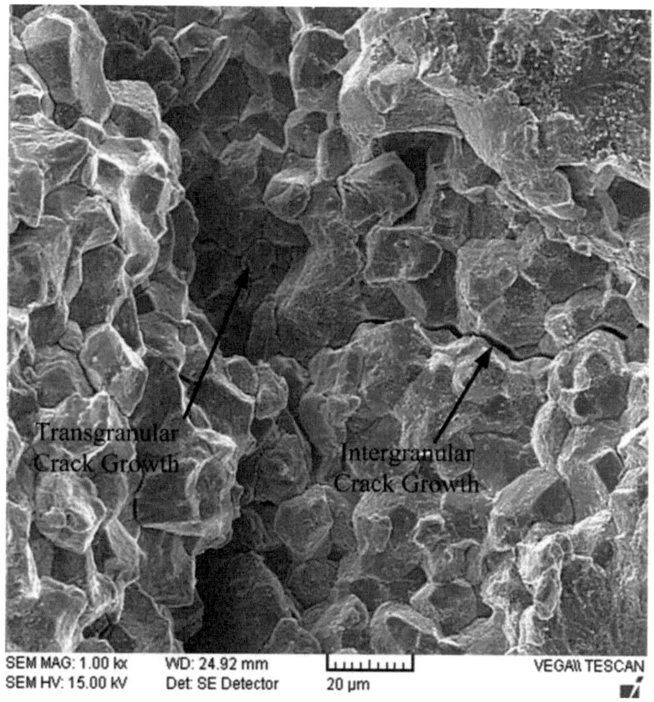

Figure 3.
Intergranular and transgranular stress corrosion cracking of the AISI 316L stainless steel at polythionic acid environment [8].

2.1 Materials

2.1.1 Stainless steels

Austenitic stainless steels suffer from SCC in chlorides, caustic, and polythionic acid. When austenitic stainless steels with sufficient carbon content (more than 0.03 wt.%) are heated in the range of 415–850°C, their microstructure becomes susceptible to precipitation of chromium carbides ($M_{23}C_6$) along grain boundaries known as sensitization [9, 12, 13]. Formation of Cr-rich carbides along grain boundaries may drastically deplete free chromium content in the area adjacent to the grain boundaries and render them susceptible to rapid preferential dissolution. Sensitized steels are most susceptible; the stress corrosion cracking of nonsensitized steels is also observed [14, 15]. Dissolution of grain boundaries in some corrosive environments aside from tensile stress led these types of materials to SCC.

2.1.2 Copper and copper alloys

Seasonal cracking of brass in the rainy season in an ammoniacal environment is another classical example of SCC. This was first identified on the brass cartridge used by the British Army in India. Since it is usually identified during the rainy season, it is also called seasonal cracking [12]. Alpha brass is an alloy of Cu-Zn. It can crack either intergranularly or transgranularly in nontarnishing ammonia solutions, depending on its zinc content [16–18]. Transgranular stress corrosion cracking, TGSCC, is observed in alloys with 20 or 30% Zn but not in alloys with 0.5 or 10% Zn [19, 20]. Stress corrosion cracking of Cu-Zn and Cu-A1 alloys in cuprous ammonia solutions can only occur when the parting limits for dealloying are exceeded. The parting limits are about 14 and 18 a/o for Cu-A1 and Cu-Zn, respectively [21]. Cu-A1 and Cu-Ga alloys have shown similar behaviors [19, 22].

2.1.3 Aluminum and aluminum alloys

Aluminum and all its alloys can fail by cracking along grain boundaries when simultaneously exposed to specific environments and stresses of sufficient magnitude [23, 24]. Of eight series of aluminum alloys, 2xxx, 5xxx, and 7xxx aluminum alloys are susceptible to SCC. Among them, 7xxx series aluminum alloys have a specific application in aerospace, military, and structural industries due to superior mechanical properties. In these high-strength 7xxx aluminum alloys, SCC plays a vital factor of consideration, as these failures are catastrophic during the service [25].

2.1.4 Carbon steels

Carbon and low alloy steels have shown SCC in a wide range of environments that tend to form a protective passive or oxide film [26–30]. The environments that would passivate carbon steels have been found to cause SCC, including strong caustic solutions, phosphates, nitrates, carbonates, ethanol, and high-temperature water. The problems are important for both economic and safety reasons, due to the extensive use of carbon steels [31]. For example, nitrate cracking in an ammonium nitrate plant caused by catastrophic failures and a lot of financial losses. Caustic cracking of steam-generating boilers made of low alloy steels was a serious problem, which led an ammonia plant to repeated emergency shutdowns.

2.1.5 Titanium alloys

Stress corrosion cracking may be a problem whenever certain high-strength titanium alloys are exposed to aqueous and certain solvent environments [32–36]. For the first time, SCC of titanium was reported by Kiefer and Harple who describe the cracking phenomena with commercially pure titanium in red fuming nitric acid [37]. Hot salt cracking of titanium alloys was reported in turbine blades that operate at high temperature in the mid-1950s. The subject became very active in the early 1960s because of the SCC problem connected to these alloys in a transportation program [38].The first known report of stress corrosion cracking of titanium alloys in room temperature aqueous environments was that of Brown. He found that titanium alloys, 8% aluminum–1% molybdenum–1% vanadium alloy (Ti, 8–1–1), were susceptible to SCC in seawater [38].

2.2 Environments

Another requirement for SCC to occur is a corrosive environment. The environments for SCC are specific because not all environments promote SCC. For those alloys that develop a protective film, an aggressive ion is required to promote SCC. The aggressive media to passive layer of stainless steels are chlorides, caustic, and polythionic acid. The austenitic stainless steel series 300 is more susceptible in an environment containing chlorides. Chlorides will not cause SCC unless an aqueous phase is present. It appears that stress corrosion cracking in austenitic stainless steels in the presence of chlorides proceeds transgranularly and usually occurs at temperature above 70°C [39, 40]. Cases of SCC due to chlorides have been experienced at ambient temperatures on parts that were subjected to heavy machining [41, 42]. Caustic embrittlement or stress corrosion cracking in caustic environment is another serious problem in austenitic stainless steels and causes many explosions and other types of failures in steam boiler and super heater components [9, 43–45]. Caustic cracking failures frequently originate in a welding area, which is intergranular, and a very concentrated caustic solution is usually necessary [40]. Polythionic acid is another environment, which causes SCC in austenitic stainless steels. Sulfur in feed gas in chemical and petrochemical plants led to formation of polythionic acid ($H_2S_xO_6$, x = 2–5), which aside from moisture also induced intergranular stress corrosion cracking in austenitic stainless steels [46, 47]. Well-known specific environments for the stress corrosion cracking in Al alloys include water vapor, aqueous solutions, organic liquids, and liquid metals [48]. The SCC of Ti alloys in aqueous chloride and methanolic chloride environments at ambient temperatures has been widely reported [49]. The summary of environments that cause SCC in mostly used alloys is presented in **Table 1**.

2.3 Stress

The stress in the form of tensile (not compressive) plays a key role in the SCC fracture processes. In fact, SCC would never have occurred in the absence of stress. The required tensile stresses may be in the form of directly applied stresses, thermal, in the form of residual stresses, or a combination of all [8, 50]:

$$\sigma = \sigma_{applied} + \sigma_{thermal} + \sigma_{residual} \tag{1}$$

For SCC to occur alone by applied stress, it must have a very high magnitude. The welding and mechanical residual stresses are the main sources of stress attributed to the stress corrosion cracking. The welding residual stress is produced as a result of nonuniform temperature changes during welding operation and can be calculated from thermal strain vectors.

The thermal strain vector, $\Delta\varepsilon^{th}$, is formulated by a temperature-dependent differential expansion coefficient (°/c) as follows [2]:

$$\{\Delta\varepsilon^{th}\} = [\alpha]\Delta T \tag{2}$$

in which $\Delta\varepsilon^{th}$ is the variation of strain, α is the thermal expansion of material, and ΔT is the temperature change.

The operational thermal stress can also be calculated from Eq. 2. Mechanical workings such as cold deformation and forming, machining, and grinding are the other sources, which introduce residual stresses [8, 51].

3. Stress corrosion cracking mechanism

Extensive investigations have been devoted to find mechanisms of SCC for different materials and environments. An SCC failure illustrates the combined effects of mechanical, physical, and chemical/electrochemical factors causing the separation of metal bonds at the crack tip, thereby advancing the crack. Three mechanisms for SCC have been proposed through the investigations [52]:

3.1 Pre-existing active path mechanism

This model supposes that there are pre-existing paths in an alloy that is susceptible to anodic dissolution. Because of precipitation or solute segregation of impurities like sulfur, phosphorus, and chromium carbides, the electrochemical properties of the matrix and segregates are changed. The area adjacent to the grain boundaries

Metal	Environment
Al alloys	$NaCl-H_2O_2$ solutions
	NaCl solutions
	Seawater
Copper alloys	Ammonia vapor and solutions
	Amines
	Water or water vapor
Gold alloys	$FeCl_3$ solutions
	Acetic acid-salt solutions
Inconel	Caustic soda solutions
Lead	Lead acetate solutions
Magnesium alloys	$NaCl-Na_2CrO_4$ solutions
	Rural and coastal atmospheres
	Seawater
	Distilled water
Nickel	Fused caustic soda
Steels	NaOH solutions
	$NaOH-Na_2SiO_4$ solutions
	Calcium, ammonium, and sodium nitrite solutions
	Mixed acids (H_2SO_4-HNO_3)
	Acidic H_2S solutions
	Seawater
	Carbonate-bicarbonate solutions
Stainless steels	Acidic chloride solutions
	$NaCl-H_2O_2$ solutions
	Seawater
	H_2S
	$NaOH-H_2S$ solutions
	Condensing steam from chloride waters
Titanium alloys	Red fuming nitric acid
	Seawater
	Methanol-HCl

Source: Craig and Lane [54].

Table 1.
Summary of some environments caused by SCC on different alloys.

is depleted from one or more alloying elements, and so under such conditions, localized galvanic cells are created (**Figure 4**). Since precipitation or segregation is generally anodic to the matrix of the grains, dissolution under an anodic reaction occurs and provides active path for localized corrosions [53]. Also, the removal of the protective film at the pre-existing crack tips by plastic deformation would facilitate the onset of localized corrosion.

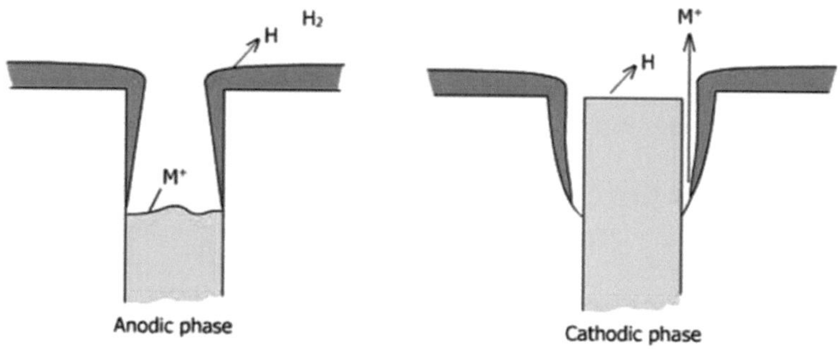

Figure 4.
Galvanic cell mechanism [52].

3.2 Strain-generated active path mechanisms

This mechanism has been extensively studied in stress corrosion cracking of alpha brass in ammoniacal environment and also proposed for caustic cracking of boiler steel. The model is based on the idea of a strain-induced rupture of the protective film, and so plastic strains play a main role in failure processes [52, 55]. The theory assumes the existence of a passivation film on a metal surface. The passivation film protects the underlying metal against corrosive agents. The passivation film is ruptured by plastic strain due to mechanical workings. After the film is ruptured, the bare metal is exposed to the corrosive environment. The processes of disruptive strain (disruption of protective film) and film formation (due to repassivation) have occurred and alternate with each other. The crack propagates when the rate of rupture of oxide film is higher than the rate of repassivation of the film [52]. The mechanism is shown in **Figure 5**.

3.3 Adsorption-related phenomenon

This model is based on the effects of environmental species on interatomic bond strength. The theoretical fracture stress required to separate two layers of atoms of spacing b is given by [56].

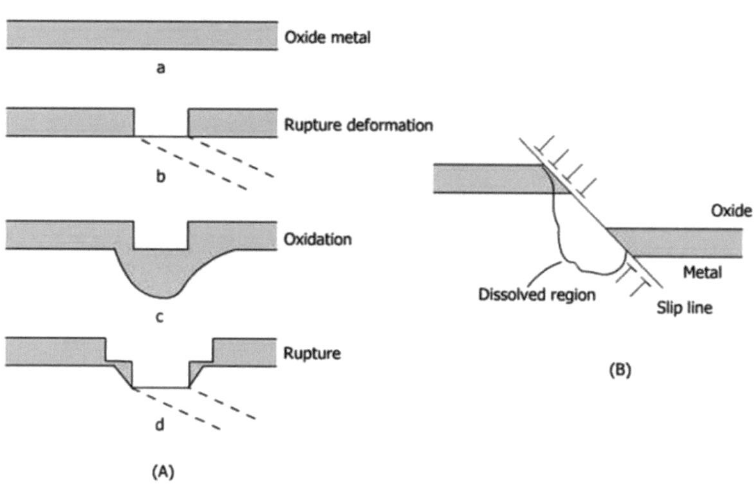

Figure 5.
Strain-generated active path mechanisms. (A) Film rupture model and (B) slip-step dissolution model [52].

$$\sigma_f = \left(\frac{E\gamma_s}{b}\right)^{1/2} \tag{3}$$

where E is the Young modulus, γ is the surface energy, and b is the spacing between atoms.

This theory implies that if surface energy is reduced, then σ_f will also be reduced. In corrosive environments, aggressive agents are present and they are absorbed at the crack tips, the surface energy is effectively lowered, and fracture takes place in stress much lower than design stress [52].

4. Application of fracture mechanics

The design of steel structure and component based on tensile properties has many disadvantages that do not take into account the role of imperfections. Fracture mechanic introduces another material characteristic, namely, fracture toughness, K_{IC}, which considered the role of cracks and imperfection in the form of cracks in designs. In its simplest form [57].

$$K_{IC} = \sigma\sqrt{\pi a} \tag{4}$$

where σ is the design stress and a is the size of an existing crack.

According to this equation, fracture occurs when stress intensity factor, K_t, on the crack tip is equal to K_{IC}. This applies for the propagation of cracks due to mechanical loadings. For the structure parts exposed to corrosive environments, the situation is quietly different. The corrosive agents caused a significant drop in the load capacity and the fracture toughness of the metals. This is typically shown in **Figure 6**. As outlined in the diagram, designs of structures in the corrosive environment based on K_{Ic} led the component to failure in a short period of time. Therefore, in these situations, K_{Ic} should be replaced with K_{Iscc}, which is the threshold value for SCC [12]. This means that in the corrosive environment, the load capacity must be significantly lower than the clean environments. The use of fracture mechanics for high-strength low alloy steels is responsive, but for austenitic steels with branch cracks, the subject should be treated with caution [12].

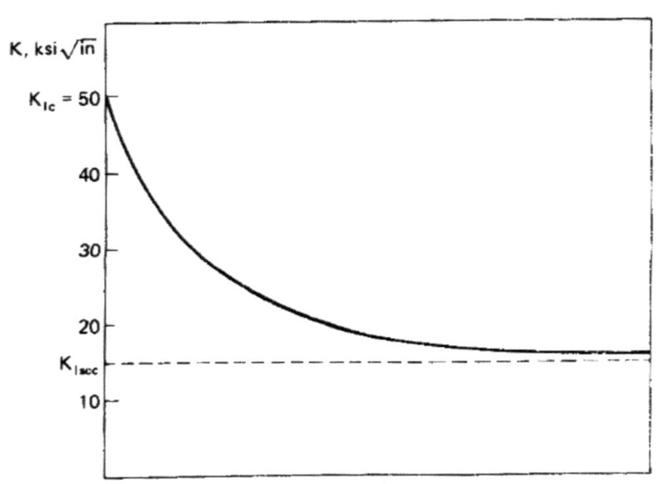

Figure 6.
Effects of corrosive environment on fracture toughness [12].

5. Prevention

Since the exact mechanism of SCC has not been completely understood, prevention methods are either general or empirical in nature. Appropriate strategy should be done in order to minimize this problem to ensure not only the safety of human life but also the safety of cost. The following general methods are recommended to overcome the SCC problems [12, 52, 58, 59]:

a. Lowering the tensile stress in the welded component using post weld heat treatment. The post weld heat treatment reduced or eliminated residual stress on surface and through the bulk of material. Plan and low alloy steels may be a stress relief at 1100–1200°F. The range of residual stress relief temperature for austenitic stainless steels is from 1500 to 1700°F. Reduction of tensile stresses by shot peening is also recommended. Shot peening introduces surface compressive stresses.

b. Eliminating aggressive agents from the environment by, for example, degasification, demineralization, or distillation.

c. Changing the alloy is one possible solution if neither the environment nor stress can be changed. For example, it is a common practice to use Inconel (raising the nickel content) when typ. 304 stainless steel is not satisfactory.

d. Applying cathodic protection: impressed current cathodic protection system has been successfully used to prevent SCC of steels.

e. Adding inhibitors to the system if feasible: high concentrations of phosphate have been successfully used.

f. Coatings are sometimes used, and they depend on keeping the environment away from the metal.

6. Failure case studies

6.1 Case 1: stress corrosion cracking of a circulation water heater tube sheet [8]

After only 3 years' service of a circulation water heater (heat exchanger), it has been shown to sever leakage and has led a methanol plant to emergency shutdown. An on-site investigation revealed extensive cracking initiated at weld area and through the tube sheet holes as it is shown in **Figure** 7.

6.1.1 Material and environment

The circulation water heater is a vertical U-type heat exchanger made of austenitic stainless steels. The equipment used to decline reforms gas temperature in a methanol plant. The hot reformed gas at approximately 385°C entered the tubes and is cooled down to 168°C by exchanging the heat with processed water in the shell. The gases that flow through the tubes are mainly CO_2, CO, H_2, CH_4, and N_2 and at a pressure of 3.9 MPa. At the shell cooling process, water flows with about 6 MPa pressure.

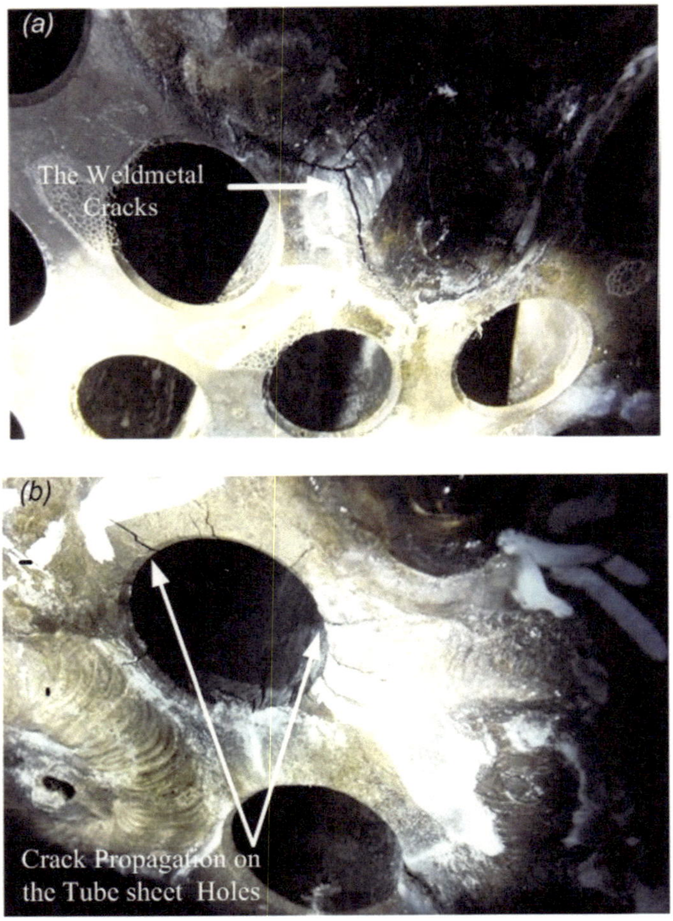

Figure 7.
Failed area (a) cracks extending in the weld joint of tube sheet to plugs and (b) branched cracks in the surface of the tube sheet and through the holes [8].

6.1.2 Cause

Deposits had formed on top of the tube sheet due to shutdown errors. AISI 316L materials overheated in service because of the insulation role of the deposits. Material sensitization occurs since overheating. The presence of sulfur in the process gas aside from moisture formed polythionic acid during shutdowns. Residual stress produced by heavy machining and welding aside from operational thermal stress provided tensile stress, which is needed for SCC. Stress corrosion cracking is induced by polythionic acid. Concentrated water with other aggressive agents such as caustic and chlorides leaked through the cracks aid the failures.

6.1.3 Prevention

- Cleaning of the shell by demineralized water after each shut down in order to prevent the forming of insulating deposits above the tube sheet
- Reduction of sulfur in feeding gas
- Reduction of caustic and chlorides in processed water

6.2 Case 2: failure of an austenitic stainless steel tubes in a gas-fired steam heater [9]

Carryover of caustic soda (NaOH) in the steam path caused catastrophic failure of superheater stainless steel tubes in a gas-fired heater and led to an unexpected shutdown after just 5 months of continuous service following the start of production. The failure areas are shown in **Figure 8**. Three types of cracks are identified in various regions of the tube: circumferential cracks adjacent to the seam weld, circumferential cracks at the ribbon of the seam weld, and longitudinal cracks on the U-bend. The path of cracks was complex on the surface or in the bulk metal; all had nucleated from inside the tubes. A visual inspection revealed a white deposit, high in sodium, around the cracks on the surface of the tubes.

6.2.1 Material and environment

The superheater tube material was made of AISI 304H austenitic stainless steel material.

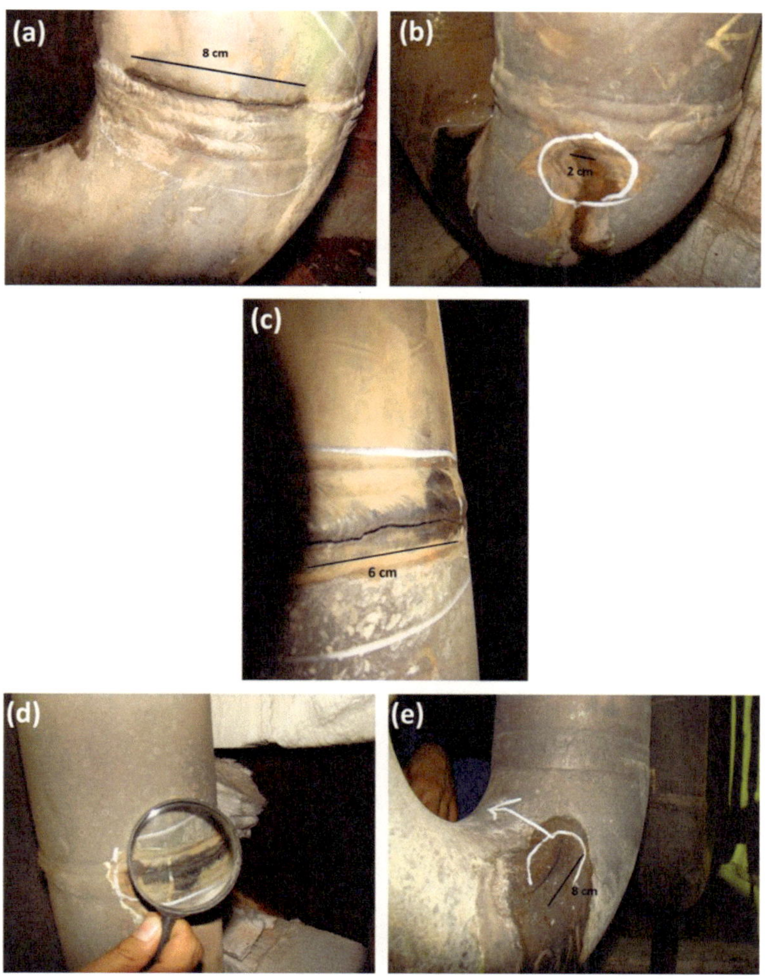

Figure 8.
(a and b) Circumferential cracks adjacent to the seam weld, (c and d) circumferential cracks at the ribbon of seam weld, and (e) longitudinal cracks on the U-bend [9].

The gas-fired steam heater (FH) generates high-pressure (HP) steam for turbines for the processing of methanol. Demineralized water for the boiler and subsequent steam path is prepared in the water treatment unit. Caustic soda is injected to demineralized water for pH control. The water is transferred to the preheat exchangers, is converted to saturated high-pressure steam at 325°C and 119 MPa, and is sent to the FH. Through the FH tubes, saturated steam converted to supersaturate steam at a temperature of 505°C and pressure of 119 MPa.

6.2.2 Cause

The main cause of crack initiation was the increase of pH due to the rise of caustic concentration in condensed drops. Sensitized austenite grains caused by chromium carbide depletion adjacent to the grain boundaries were attacked by concentrated caustic in the HAZ metal and U-bend area and led the heater to the caustic SCC failure.

6.2.3 Prevention

a. Using A335 Grade P9, a low alloy steel tube shows higher resistance to SCC than AISI 304H stainless steel

b. Proper discharge of the tubes during shutdowns to prevent the formation of the concentrated deposits of caustic through the tubes

6.3 Case 3: failures of brass condenser tubes [60]

After a general overhaul of a thermal power plant in Serbia in November 2014, failure of hundreds of brass condenser tubes occurred during the hydrostatic test. Also, it was noted that some backing plates had fallen off from the tubes before this test. Fracture is observed only in condenser tubes of brass, as can be seen in **Figure 9**.

6.3.1 Material and environment

The failed tube material of the condenser was made of brass CuZn28Sn1 (admiralty brass). The cooling water (roughly filtered river water) flows through the tubes, while the hot steam flows around the tubes.

Figure 9.
Failure of brass condenser tubes near joining location with backing plate.

6.3.2 Cause

Analysis of fracture surfaces using scanning electron microscopy (SEM) has shown the brittle transgranular fracture due to the occurrence of SCC. The condenser tubes are made of brass CuZn28Sn1. Ammonia and other nitrogen compounds in the cooling water through the tubes were found. These compounds are specific agents that cause stress corrosion cracking (SCC) in brass. In the joining region of condenser tubes to backing plates, there are residual tensile stresses. During the floods in May 2014, there was an increase in the concentration of ammonia and other nitrogen compounds in the river cooling water flowing through the condenser tubes. Failure of brass condenser tubes occurred due to SCC, because the necessary conditions for the SCC occurrence were fulfilled.

6.3.3 Prevention

- The risk of SCC in brass condenser tubes can be reduced if specific substances responsible for SCC occurrence are removed, as much as possible. This can be achieved by cleaning and drying the tubes immediately after the operation delay of the power plant.

- Another way to reduce the risk of SCC occurrence in condenser tubes is the replacement of existing tubes (made of brass CuZn28Sn1, very susceptible to SCC) with tubes made of alloys of greater resistance to SCC, such as copper-nickel alloys or Bi-brass alloys [61].

7. Conclusion

Stress corrosion cracking is one of the main causes of unforeseen and dangerous destruction of industrial plants. The sensitized material, certain environments, and stress are three factors necessary for the occurrence of these types of failures. The environment prone to the cracking for each metal or alloy is specific because not all environments promote the SCC. Austenitic stainless steels suffer from SCC in chlorides, caustic, and polythionic acid. Copper alloys corrode in ammonia-containing environments. Well-known specific environments for the stress corrosion cracking in Al alloys include water vapor, aqueous solutions, organic liquids, and liquid metals. The SCC of Ti alloys in aqueous chloride and methanolic chloride environments has been widely reported. The tensile stress plays a key role in the stress corrosion cracking phenomenon. The required tensile stresses may be in the form of directly applied stresses, thermal, in the form of residual stresses, or a combination of all.

If one of these three components does not exist, this type of corrosion will not occur. Therefore, the solving methods should be based on the elimination of one of these three factors. Corrosive environment modification, the stress in the form of compression, and using proper material are three general proposed methods of prevention.

Author details

Alireza Khalifeh
Department of Materials Science and Engineering, School of Engineering, Shiraz University, Shiraz, Iran

*Address all correspondence to: areza1006@gmail.com; a.khalifeh@shirazu.ac.ir

References

[1] Abedi SS, Abdolmaleki A, Adibi N. Failure analysis of SCC and SRB induced cracking of a transmission oil products pipeline. Engineering Failure Analysis. 2007;**14**(1):250-261

[2] Xu S, Zhao Y. Using FEM to determine the thermo-mechanical stress in tube to tube-sheet joint for the SCC failure analysis. Engineering Failure Analysis. 2013;**34**:24-34

[3] Kumar MS et al. Failure analysis of a stainless steel pipeline. Engineering Failure Analysis. 2008;**15**(5):497-504

[4] Meresht ES, Farahani TS, Neshati J. Failure analysis of stress corrosion cracking occurred in a gas transmission steel pipeline. Engineering Failure Analysis. 2011; **18**(3):963-970

[5] Rao T, Nair K. Microbiologically influenced stress corrosion cracking failure of admiralty brass condenser tubes in a nuclear power plant cooled by freshwater. Corrosion Science. 1998; **40**(11):1821-1836

[6] Prawoto Y et al. Effect of microstructures on SCC of steel: Field failure analysis case study and laboratory test result. Engineering Failure Analysis. 2011;**18**(7):1858-1866

[7] Swaminathan J et al. Sensitization induced stress corrosion failure of AISI 347 stainless steel fractionator furnace tubes. Engineering Failure Analysis. 2011;**18**(8):2211-2221

[8] Khalifeh A et al. Stress corrosion cracking of a circulation water heater tubesheet. Engineering Failure Analysis. 2017;**78**:55-66

[9] Parnian N. Failure analysis of austenitic stainless steel tubes in a gas fired steam heater. Materials & Design. 2012;**36**:788-795

[10] Vinoy T et al. Stress corrosion crack growth studies on AISI typ. 316 stainless steel in boiling acidified sodium chloride solution. Journal of Nuclear Materials. 1996;**238**(2): 278-284

[11] Li Y et al. Failure analysis of the 304 stainless steel tube in a gas analyzer. Engineering Failure Analysis. 2012;**20**: 35-42

[12] Fontana MG. Corrosion Engineering. India: Tata McGraw-Hill Education; 2005

[13] Sedriks AJ. Corrosion of Stainless Steel, 2nd ed. USA: Wiley-Interscience; 1996

[14] Pal S, Ibrahim R, Raman RS. Studying the effect of sensitization on the threshold stress intensity and crack growth for chloride stress corrosion cracking of austenitic stainless steel using circumferential notch tensile technique. Engineering Fracture Mechanics. 2012;**82**:158-171

[15] Angeliu T et al. Intergranular stress corrosion cracking of unsensitized stainless steels in BWR environments. In: Ninth International Symposium on Environmental Degradation of Materials in Nuclear Power Systems-Water Reactors. Wiley Online Library; 1999

[16] Davis JR. Copper and Copper Alloys. USA: ASM International; 2001

[17] El-Amoush AS et al. Stress corrosion cracking of the pre-immersed tin brass heat exchanger tube in an ammoniacal solution. Materials & Design (1980–2015). 2014;**56**:842-847

[18] Kannan MB, Shukla P. Stress corrosion cracking (SCC) of copper and copper-based alloys. In: Stress Corrosion Cracking. USA: Elsevier; 2011. pp. 409-426

[19] Craig J, Montague W, Pugh E. Factors influencing the path of stress-corrosioncracking in alpha-phase copper alloys exposed to aqueous ammonia environments. ASM Transactions Quarterly. 1968;**61**(3):468-473

[20] Pugh, E., Mechanisms of Stress Corrosion Cracking of Alpha-Brass in Aqueous Ammonia. 1971.

[21] Sieradzki K et al. The relationship between dealloying and transgranular stress-corrosion cracking of Cu-Zn and Cu-Al alloys. Journal of the Electrochemical Society. 1987;**134**(7): 1635-1639

[22] Ohtani N, Dodd R. Mechanisms of transgranular stress corrosion cracking of solid solution alloys. Corrosion. 1965; **21**(5):161-172

[23] Popović M, Romhanji E. Stress corrosion cracking susceptibility of Al-Mg alloy sheet with high Mg content. Journal of Materials Processing Technology. 2002;**125**:275-280

[24] Jones R et al. Role of Mg in the stress corrosion cracking of an Al-Mg alloy. Metallurgical and Materials Transactions A. 2001;**32**(7):1699-1711

[25] Rao AU et al. Stress corrosion cracking behaviour of 7xxx aluminum alloys: A literature review. Transactions of Nonferrous Metals Society of China. 2016;**26**(6):1447-1471

[26] Sridhar N et al. Stress corrosion cracking of carbon steel in ethanol. Corrosion. 2006;**62**(8):687-702

[27] Wang J, Atrens A. SCC initiation for X65 pipeline steel in the "high" pH carbonate/bicarbonate solution. Corrosion Science. 2003;**45**(10): 2199-2217

[28] Cheng Y, Steward F. Corrosion of carbon steels in high-temperature water studied by electrochemical techniques. Corrosion Science. 2004;**46**(10): 2405-2420

[29] Contreras A et al. Mechanical and environmental effects on stress corrosion cracking of low carbon pipeline steel in a soil solution. Materials & Design. 2012;**35**:281-289

[30] Jones RH. Stress-Corrosion Cracking, Materials Performance and Evaluation. USA: ASM International; 2017

[31] Cheng YF. Stress Corrosion Cracking of Pipelines. Vol. 15. USA: John Wiley & Sons; 2013

[32] Zardiackas L, Bogan J. Stress corrosion cracking resistance of titanium implant materials. In: 27th Annual Meeting of the Society for Biomaterials in Conjunction with the 33rd International Biomaterials Symposium. 2001

[33] Wang J et al. Stress corrosion cracking of NiTi in artificial saliva. Dental Materials. 2007;**23**(2):133-137

[34] Guo X et al. Correlation between passive film-induced stress and stress corrosion cracking of α-Ti in a methanol solution at various potentials. Materials Science and Engineering A. 2003;**346** (1–2):1-7

[35] Trasatti S, Sivieri E. Electrochemical and stress corrosion cracking behaviour of titanium in n-propanol and iso-propanol solutions. Materials Chemistry and Physics. 2004;**83**(2–3):367-372

[36] Pilchak A, Young A, Williams J. Stress corrosion cracking facet crystallography of Ti–8Al–1Mo–1V. Corrosion Science. 2010;**52**(10): 3287-3296

[37] Kiefer GC, Harple WW. Stress-corrosion cracking of commercially pure

titanium. Metal Progress. 1953;**63**(2): 74-76

[38] Blackburn M, Feeney J, Beck T. Stress-corrosion cracking of titanium alloys. In: Advances in Corrosion Science and Technology. USA: Springer; 1973. pp. 67-292

[39] Kan W, Pan H. Failure analysis of a stainless steel hydrotreating reactor. Engineering Failure Analysis. 2011; **18**(1):110-116

[40] Shipley RJ, Becker WT, editors. Handbook ASM: Failure Analysis and Prevention. Vol. 11. USA: ASM International; 2002

[41] Fairweather N, Platts N, Tice D. Stress-corrosion crack initiation of typ. 304 stainless steel in atmospheric environments containing chloride: Influence of surface condition, relative humidity, temperature and thermal sensitization. In: Corrosion 2008. USA: NACE International; 2008

[42] Hayashibara H, Mayuzumi M, Mizutani Y. Effects of temperature and humidity on atmospheric stress corrosion cracking of 304 stainless steel. In: Corrosion 2008. USA: NACE International; 2008

[43] Wang Y, Lu Y-B, Pan H-L. Failure analysis of a hydro-processing reactor. Engineering Failure Analysis. 2009; **16**(1):11-18

[44] Raman RS. Role of caustic concentration and electrochemical potentials in caustic cracking of steels. Materials Science and Engineering A. 2006;**441**(1):342-348

[45] Abouswa K, Elshawesh F, Abuargoub A. Stress corrosion cracking (caustic embrittlement) of super heater tubes. Desalination. 2008;**222**(1): 682-688

[46] Singh PM, Ige O, Mahmood J. Stress corrosion cracking of typ. 304L stainless steel in sodium sulfide-containing caustic solutions. Corrosion. 2003; **59**(10):843-850

[47] Horowitz HH. Chemical studies of polythionic acid stress-corrosion cracking. Corrosion Science. 1983;**23**(4): 353-362

[48] Speidel MO, Hyatt MV. Stress-corrosion cracking of high-strength aluminum alloys. In: Advances in Corrosion Science and Technology. USA: Springer; 1972. pp. 115-335

[49] Staehle RW. Fundamental Aspects of Stress Corrosion Cracking: Proceedings of Conference. National Association of Corrosion Engineers; 1969

[50] Ghosh S et al. Role of residual stresses induced by industrial fabrication on stress corrosion cracking susceptibility of austenitic stainless steel. Materials & Design. 2011;**32**(7): 3823-3831

[51] Turnbull A et al. Sensitivity of stress corrosion cracking of stainless steel to surface machining and grinding procedure. Corrosion Science. 2011; **53**(10):3398-3415

[52] Ahmad Z. Principles of Corrosion Engineering and Corrosion Control. UK: Butterworth-Heinemann; 2006

[53] Hoar T, Hines J. In: Robertson WD, editor. Stress Corrosion Cracking and Hydrogen Embrittlement. Vol. 107. New York: John Wiley & Sons; 1956

[54] Craig B, Lane R. Environmentally-assisted cracking: Comparing the influence of hydrogen, stress, and corrosion on cracking mechanisms. AMPTIAC Quarterly. 2005;**9**(1):17-24

[55] Champion F. Symposium on Internal Stresses in Metals and Alloys. London: Institute of Metals; 1948. p. 468

[56] Murgatroyd J, Sykes R. Mechanism of brittle rupture. Nature. 1945; **156**(3972):716

[57] Dieter GE, Bacon DJ. Mechanical Metallurgy. Vol. 3. New York: McGraw-Hill; 1986

[58] Du X et al. Inhibitive effects and mechanism of phosphates on the stress corrosion cracking of brass in ammonia solutions. Corrosion Science. 2012;**60**: 69-75

[59] Kannan MB, Raja V. Enhancing stress corrosion cracking resistance in Al-Zn-Mg-Cu-Zr alloy through inhibiting recrystallization. Engineering Fracture Mechanics. 2010;**77**(2):249-256

[60] Bobić B et al. Failures of Brass Condenser Tubes. Integritet I Vek Konstrukcija. 2016;**16**(1):19-23

[61] You S-J et al. Stress corrosion cracking properties of environmentally friendly unleaded brasses containing bismuth in Mattsson's solution. Materials Science and Engineering A. 2003;**345**(1–2):207-214

Chapter 4

Micromechanical Failure Analysis of Unidirectional Composites

Zheng-Ming Huang

Abstract

Internal stresses in the fiber and matrix of a unidirectional (UD) composite obtained by any micromechanics model are homogenized quantities. They must be converted into true values before an effective specifically failure and strength property of the composite can be predicted in terms of the fiber and matrix properties only. As elastic property of a material does not depend on the magnitude of its stresses, the predictions of an elastic property of the composite based on the homogenized and true stresses of the constituents are the same, concealing the fact that the elastic property should be predicted based on the true stresses as well. The conversion of all of the internal stress components has been shown in this chapter. Predictability of a total number of 12 micromechanics models for the stiffness and strength of a UD composite is assessed against the experimental data of the 9 UD composites provided in three worldwide failure exercises (WWFEs). Bridging Model exhibits overall the best accuracy in both the stiffness and the strength predictions. Further, the smallest fiber volume in a RVE (representative volume element) for an FE (finite element) approach plays a much more dominant role than other issues such as a random fiber arrangement pattern to achieve the highest simulation accuracy. Finally, consistency of a micromechanics model in calculating the internal stresses of a composite is an issue that should be taken into account. Only Bridging Model is consistent. A non-consistency implies that a full three-dimensional (3D) model should be used to predict an effective property, e.g., failure behavior of a composite even though it is only subjected to a uniaxial load, and a 3D RVE geometry should be discretized if a numerical micromechanics approach is applied.

Keywords: composites, micromechanics, stress concentration factors, interface crack, failure analysis, strength prediction, consistency, fiber arrangement arrays

1. Introduction

Fiber-reinforced composites have been used as a primary-load carrying structural material in many engineering areas especially in aerospace industry. Due to their anisotropy, the mechanical properties of the composites are difficult or expensive to understand through experiments. Establishment of mathematical models to link the overall behaviors of the composites with their constituent structures and properties is an objective of micromechanics. As any continuous fiber-reinforced composite structure can be subdivided into a combination of a series of unit cells or RVEs [1], which can be considered as UD composites in their local coordinate system, a micromechanics analysis of a UD composite is fundamental.

So far, numerous micromechanical models have been developed to predict elastic properties of the composites from those of the constituent fiber and matrix materials [2–4]. On the other hand, very few of them have been applied to estimate failure and strength behaviors of the composites only based on the original constituent data measured independently with a reasonable accuracy [5]. This is attributed to that the internal stresses in the fiber and matrix of a composite obtained by a micromechanics theory are homogeneous quantities. They must be converted into true values before an effective property of the composite is predicted in terms of the original consituent properties. An elastic property (modulus, stiffness, etc.) of a material does not depend on the magnitude of the stresses in it, as long as they do not exceed the elastic limit of the material. Hence, the predictions of an elastic property of the composite based on the homogenized and true stresses of the constituents are the same, concealing the fact that the composite elastic property should also be predcited based on the true stresses. The stress field of the fiber is uniform [6, 7]. Its homogenized and true stresses are the same. A true stress of the matrix is obtained by multiplying its homogeneous counterpart with a stress concentration factor (SCF) of the matrix in the composite. This is because a plate with a hole generates a stress concentration if subjected to an in-plane tension. When the hole is filled with a fiber of different proerties, a stress concentration occurs as well.

The most significant feature is that such an SCF cannot be defined, following a classical approach, as a maximum point-wise stress divided by the overall applied one. Otherwise, the resulting SCF would be infinite if there is an interface crack or debonding on a fiber and matrix interface, since at the crack tip, a matrix stress is singular. All of the SCFs of the matrix in a composite have been obtained [8–11] and are summarized in this chapter.

Another objective of this chapter is to make a critical assessment for the predicability of 12 well-known micromechanical models for the stiffness and mainly failure and strength of a UD composite, based on the original fiber and matrix properties and the fiber volume fraction. By "original," it is meant that the properties are either measured independently using monolithic material, e.g., matrix specimens or documented in a recognized material database. The previous comparisons, e.g., Refs. [12–14], were made essentially for the stiffness predictions by different models. Seldom have been found for the strength predictions. The models considered in this chapter are Eshelby's method [6, 15], Bridging Model [16], Mori-Tanaka method [17, 18], rule of mixture method [19], Chamis model [20], modified rule of mixture method [19], Halpin-Tsai formulae [21], Hill-Hashin-Christensen-Lo model [22], self-consistent method [15], generalized self-consistent method [15, 23], double inclusion method [4], and finite element method (FEM) [24] with different fiber arrangement patterns in a unit cell or RVE. The measured stiffness and strength data of all of the nine independent UD composites adopted in three WWFEs [25–27] are used as benchmark to judge the prediction accuracy of each model. An accuracy ranking is made based on the overall correlation errors between the models' predictions and the experiments.

Numerical micromechanics such as FE approaches are popular in the current literature. What kind of fiber arrangement pattern should be chosen in a RVE? By definition, the RVE geometry, on which a homogenization is made, should be infinitesimal. In reality, however, a finite volume for a RVE has to be used. Thus, various choices exist. Many people [28–30] deemed that a random arrangement pattern with quite a number of fibers, e.g., 30 [31], 40 [32], or even 120 [33], should be contained in the RVE. It is shown among the four different patterns considered in this chapter the one with the smallest fiber volume results in the highest prediction accuracy.

Finally, consistency in the internal stress calculation deserves an attention. Any micromechanics model can result in two sets of formulae, i.e., two-dimensional (2D) and three-dimensional (3D) formulae, for homogenized internal stresses in the fiber and matrix of a composite. When the composite is subjected to a planar load, either the 2D or the 3D formulae can be applied to calculate the internal stresses. If the stress components in the fiber and matrix by the 2D and the 3D formulae are exactly the same, the model is said to be consistent in the internal stress calculation. It is shown in the chapter that among the 12 theories considered, only Bridging Model is consistent. Thus, if any other models specifically the numerical micromechanics method is applied to determine the internal stresses, its full 3D formulae should be made in use, even though the composite is subjected to only a uniaxial load.

In short, the topics addressed in this chapter are important to the micromechanical analysis of an effective mechanical property of a composite. If one would like to estimate its failure and strength behavior under an arbitrary load condition only from its constituent information, the true stress theory as described in this chapter is inevitable.

2. Fundamental for internal stresses

A composite is heterogenous by nature. Any stress and strain should be defined upon averaged quantities with respect to its RVE of a volume V' through.

$$\sigma_i = \left(\int_{V'} \tilde{\sigma}_i dV \right) / V' = V_f \sigma_i^f + V_m \sigma_i^m, \tag{1}$$

$$\varepsilon_i = \left(\int_{V'} \tilde{\varepsilon}_i dV \right) / V' = V_f \varepsilon_i^f + V_m \varepsilon_i^m. \tag{2}$$

It must be realized that by definition V' should be infinitesimal, and a resulting stress or strain, with $\sim$ on head, represents a point-wise quantity. If V' is finite, the corresponding one is called a homogenized quantity. In Eqs. (1) and (2), V is a volume fraction with $V_f + V_m = 1$. A super-/subscript f or m refers to the fiber or matrix, whereas a quantity without any suffix is related to the composite.

Using a bridging equation, $\{\sigma_i^m\} = [A_{ij}]\{\sigma_j^f\}$, and the constitutive relationships, $\{\varepsilon_i^f\} = [S_{ij}^f]\{\sigma_j^f\}$, $\{\varepsilon_i^m\} = [S_{ij}^m]\{\sigma_j^m\}$, and $\{\varepsilon_i\} = [S_{ij}]\{\sigma_j\}$, the internal stresses in the fiber and matrix together with the compliance tensor of the composite are found to be [16]

$$\{\sigma_i^f\} = (V_f[I] + V_m[A_{ij}])^{-1}\{\sigma_j\}, \tag{3}$$

$$\{\sigma_i^m\} = [A_{ij}](V_f[I] + V_m[A_{ij}])^{-1}\{\sigma_j\}. \tag{4}$$

$$[S_{ij}] = \left(V_f[S_{ij}^f] + V_m[S_{ij}^m][A_{ij}]\right)(V_f[I] + V_m[A_{ij}])^{-1}. \tag{5}$$

$\left[S_{ij}^{f}\right]$ and $\left[S_{ij}^{m}\right]$ are the compliance tensors of the fiber and matrix, respectively, and $[I]$ is a unit tensor. From Eq. (5), the bridging tensor is solved as

$$[A_{ij}] = V_f\left([S_{ij}] - \left[S_{ij}^{m}\right]\right)^{-1}\left(\left[S_{ij}^{f}\right] - [S_{ij}]\right)/V_m. \tag{6}$$

The prediction of elastic moduli is equivalent to the calculation of internal stresses in the fiber and matrix of the same composite.

3. Highlight on micromechanics models

A different micromechanics model corresponds to a different bridging tensor. Perhaps the most compact bridging tensor is given by Bridging Model [16], of which the non-zero bridging tensor elements are expressed as follows:

$$a_{11} = E^m/E_{11}^f, \tag{7.1}$$

$$a_{12} = a_{13} = \frac{\nu^m E_{11}^f - E^m \nu_{12}^f}{E^m - E_{11}^f}(a_{11} - a_{22}), \tag{7.2}$$

$$a_{22} = a_{33} = a_{44} = 0.3 + 0.7\frac{E^m}{E_{22}^f}, \tag{7.3}$$

$$a_{55} = a_{66} = 0.3 + 0.7\frac{G^m}{G_{12}^f}. \tag{7.4}$$

E_{11}^f, E_{22}^f, and G_{12}^f are longitudinal, transverse, and in-plane shear moduli of the fiber, respectively. ν_{12}^f is its longitudinal Poisson's ratio. E^m and G^m are Young's and shear moduli of the matrix. Substituting the so-defined bridging tensor into Eqs. (3) and (4) leads to

$$\sigma_{11}^f = \frac{\sigma_{11}^0}{V_f + V_m a_{11}} - \frac{V_m a_{12}\left(\sigma_{22}^0 + \sigma_{33}^0\right)}{\left(V_f + V_m a_{11}\right)\left(V_f + V_m a_{22}\right)}, \tag{8.1}$$

$$\sigma_{11}^m = \frac{a_{11}\sigma_{11}^0}{V_f + V_m a_{11}} + \frac{V_f a_{12}\left(\sigma_{22}^0 + \sigma_{33}^0\right)}{\left(V_f + V_m a_{11}\right)\left(V_f + V_m a_{22}\right)}, \tag{8.2}$$

$$\sigma_{ij}^f = \frac{\sigma_{ij}^0}{V_f + V_m a_{22}}, \; ij = 22, 33, \text{ and } 23, \tag{8.3}$$

$$\sigma_{ij}^m = \frac{a_{22}\sigma_{ij}^0}{V_f + V_m a_{22}}, \; ij = 22, 33, \text{ and } 23, \tag{8.4}$$

$$\sigma_{ij}^f = \frac{\sigma_{ij}^0}{V_f + V_m a_{66}}, \; ij = 12 \text{ and } 13, \tag{8.5}$$

$$\sigma_{ij}^m = \frac{a_{66}\sigma_{ij}^0}{V_f + V_m a_{66}}, \; ij = 12 \text{ and } 13, \tag{8.6}$$

$\{\sigma_{11}^0, \sigma_{22}^0, \sigma_{33}^0, \sigma_{23}^0, \sigma_{13}^0, \text{ and } \sigma_{12}^0\}$ are any arbitrary loads applied on the composite.

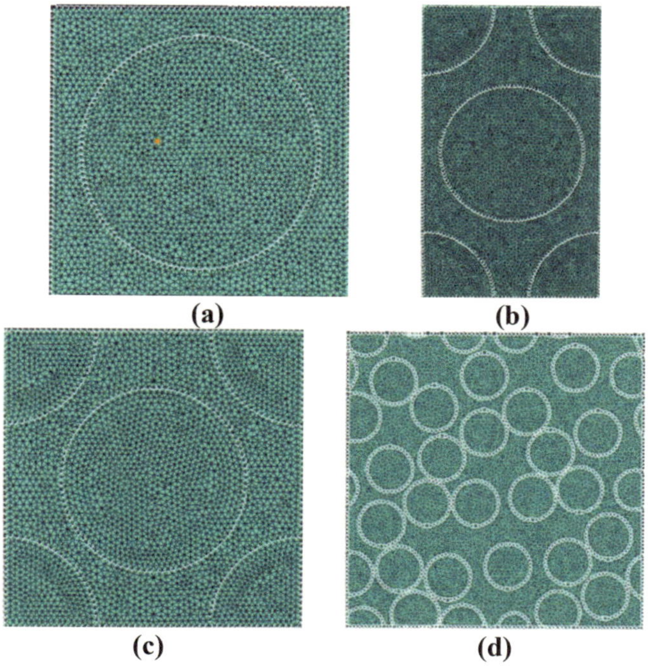

Figure 1.
Different RVEs for a UD composite used in FE approach: (a) square fiber array, (b) hexagonal fiber array, (c) square-diagonal fiber array, and (d) random fiber array.

The other analytical models are summarized in Appendix A. However, numerical micromechanics methods are even more widely applied in the current literature [34–36]. Among, an FE approach is the most common. In this approach, the fiber and matrix in a RVE geometry are discretized, respectively, into a number of elements with prescribed boundary conditions. After the stresses (i.e., point-wise quantities) in the fiber and matrix of the RVE under any load are obtained through an FE package such as ABAQUS, they are homogenized as per Eq. (1) and the bridging tensor is determined following a method of Ref. [37]. The specification of the boundary conditions has become standard [36]. Thus, different solutions only come from different RVE geometries used.

In this chapter, four kinds of RVE geometries with the same boundary conditions are chosen for the comparison. They are square fiber array [38] (**Figure 1a**), hexagonal array [36] (**Figure 1b**), square-diagonal array [39] (**Figure 1c**), and random array with 30 fibers involved [28, 31] (**Figure 1d**). Our solutions are the same as those in Ref. [36] for **Figure 1b**, in Ref. [39] for **Figure 1c**, and in Ref. [31] for **Figure 1d**, respectively.

4. Assessment on stiffness prediction

Hinton et al. organized three WWFEs to judge efficiency of the current theories for composites [5]. A total number of nine independent material systems were used. Mechanical properties of the fibers and matrices as well as fiber volume fractions of the nine UD composites were provided [25–27] and cited in **Table B.1** (see Appendix B). Measured effective properties of the composites from the exercise organizers [25–27], which are used as a benchmark to assess the predictability of the 12 models, are listed in **Table B.2**. Predictions for the five effective elastic moduli of

Model	N	Averaged error*	Error ratio	Rank	Model	N	Averaged error*	Error ratio	Rank
Bridging model	45	10.38%	1.0	1	Halpin-Tsai formulae	45	19.24%	1.85	9
FE-square	45	13.08%	1.26	2	Modified rule of mixture	45	19.35%	1.86	10
Double inclusion method	45	13.6%	1.31	3	Mori-Tanaka method	45	19.59%	1.89	11
Chamis model	45	14.09%	1.36	4	FE-square diagonal	45	21.48%	2.07	12
Hill-Hashin-C-L model	33	17.22%	1.66	5	Self-consistent method	45	21.82%	2.1	13
FE-random	45	17.57%	1.69	6	Rule of mixture method	45	28.4%	2.74	14
Generalized self-consistent	45	18.14%	1.75	7	Eshelby's method	45	30.72%	2.96	15
FE-hexagonal	45	19.05%	1.84	8					

*$=\frac{1}{N}\sum_{i=1}^{N} abs(error)_i$.

Table 1.
Overall averaged errors in prediction of the elastic moduli of the nine UD composites by different models.

each of the 9 composites by the 12 models are made and are summarized in **Table B.3**. Relative error of each predicted result in comparison with the measured counterpart (**Table B.2**) is calculated. The overall averaged errors by the 12 models are indicated in **Table 1**, in which FE-square, FE-hexagonal, FE-square-diagonal, and FE-random stand for the FEM solutions based on **Figure 1a–d**, respectively.

It is seen from the table that Bridging Model exhibits overall the highest accuracy in the stiffness prediction, with an overall correlation error of 10.48%. The second smallest error, 13.06%, is achieved by the FE-square array. The FE approach with a random fiber array of 30 fibers results in a correlation error of 17.57%, which is 34.5% less accurate than the FE approach with the square fiber array. The other two fiber arrangement patterns, the hexagonal and the square-diagonal fiber arrays, make the correlation even poorer than the random fiber arrangement. Although the four kinds of fiber arrangement patterns considered in this work may be limited, compared to unlimited possibilities in fiber arrangements, the present study confirms that the minimum fiber volume in the RVE geometry for a composite is the most dominant factor to influence the simulation accuracy, as long as suitable boundary conditions have been equally specified. **Table 1** also indicates that three analytical micromechanics models, Bridging Model, double inclusion method, and Chamis model, possess sufficient accuracy in modeling of composite stiffness.

The largest correlation error, 30.7%, is assumed by Eshelby's method. In addition to it, there are three other models attaining an averaged correlation error of more than 20% in the stiffness prediction. They are the FE-square diagonal fiber array, self-consistent method, and rule of mixture method.

5. SCFs of the matrix in a composite

5.1 Background

Let the E-glass/LY556 UD composite in **Table B.1** be subjected to only a transverse tension, σ_{22}^0, which will fail from a matrix failure. The only non-zero internal stresses of the matrix from Eqs. (8.2), (8.4), and (8.6) are $\sigma_{11}^m = 0.134\sigma_{22}^0$ and $\sigma_{22}^m = 0.442\sigma_{22}^0$. Thus, the transverse tensile strength of the composite is $\sigma_{22}^{u,t}$ = $Y_m/0.422$, where Y_m is the in situ transverse tensile strength of the matrix in the composite. Setting Y_m = $\sigma_{u,t}^m$ = 80 MPa (**Table B.2**), where $\sigma_{u,t}^m$ is the original tensile strength of the matrix, one obtains $\sigma_{22}^{u,t}$ = 181 MPa, which is more than 5.2 times greater than 35 MPa, the measured counterpart of the composite (**Table B.2**). A similar conclusion can be drawn no matter which other composite is considered or another micromechanics theory is employed. This implies that the homogenized internal stresses evaluated through Eqs. (3) and (4) must be converted into "true" values before a failure assessment can be made against the original strength data of the constituents. As point-wise strains in the fiber are uniform [6], its homogenized and true stresses are the same. However, those in the matrix are not. Each of its true stresses is obtained by multiplying the homogenized counterpart with a factor, which is agreed to call an SCF of the matrix in the composite.

5.2 Definition

The most significant feature, as aforementioned, is that such an SCF is no longer obtainable from a classical approach. Thus, the new definition must be made on an averaged stress. But with respect to which kind of geometry the averaging should be performed? A classical SCF was obtained by a point-wise (something like zero-dimensional) stress divided by an overall applied one, which is in fact a 2D quantity averaged with respect to the boundary surface. By similarity, a present SCF must be defined as a line-averaged (one-dimensional) stress of the matrix divided by a volume-averaged (3D) one since three is the maximum attainable dimension in the denominator. An SCF of the matrix subjected to a transverse load is derived through [10]

$$K_{22}(\varphi) = \frac{1}{\left|\vec{R}_\varphi^b - \vec{R}_\varphi^a\right|} \int_{\left|\vec{R}_\varphi^a\right|}^{\left|\vec{R}_\varphi^b\right|} \frac{\tilde{\sigma}_{22}^m}{\left(\sigma_{22}^m\right)_{BM}} d\left|\vec{R}_\varphi\right|, \tag{9}$$

in which $\tilde{\sigma}_{22}^m$ is a point-wise stress of the matrix determined on a concentric cylinder assemblage (CCA) model along the loading direction; $\left(\sigma_{22}^m\right)_{BM}$ is given by Bridging Model, i.e., by Eq. (8.4), φ is the inclined angle of the outward normal to a failure surface under the given load, and $\vec{R}_\varphi^a$ and $\vec{R}_\varphi^b$ are the vectors of $\vec{R}_\varphi$ at the surfaces of the fiber and matrix cylinders within the RVE, respectively, where $b = a/\sqrt{V_f}$.

5.3 Transverse SCFs

In such a load case, the explicit integration of Eq. (9) leads to [8–10]

$$K_{22}(\varphi) = \left\{1 + \frac{A}{2}\sqrt{V_f}\cos 2\varphi + \frac{B}{2\left(1 - \sqrt{V_f}\right)}\left[V_f^2 \cos 4\varphi + 4V_f \cos^2\varphi(1 - 2\cos 2\varphi) + \sqrt{V_f}(2\cos 2\varphi + \cos 4\varphi)\right]\right\}(V_f + a_{22}V_m)/a_{22}, \tag{10.1}$$

$$A = \frac{2E_{22}^{f}E^{m}\left(\nu_{12}^{f}\right)^{2} + E_{11}^{f}\left\{E^{m}\left(\nu_{23}^{f}-1\right) - E_{22}^{f}\left[2(\nu^{m})^{2}+\nu^{m}-1\right]\right\}}{E_{11}^{f}\left[E_{22}^{f}+E^{m}\left(1-\nu_{23}^{f}\right)+E_{22}^{f}\nu^{m}\right] - 2E_{22}^{f}E^{m}\left(\nu_{12}^{f}\right)^{2}}, \tag{10.2}$$

$$B = \frac{E^{m}\left(1+\nu_{23}^{f}\right) - E_{22}^{f}(1+\nu^{m})}{E_{22}^{f}\left[\nu^{m}+4(\nu^{m})^{2}-3\right] - E^{m}\left(1+\nu_{23}^{f}\right)}. \tag{10.3}$$

Under a transverse tension, the failure surface of the composite is perpendicular to the loading and hence $\varphi = 0$ (**Figure 2a**). When a transverse compression is applied, the failure surface of the composite has an inclined angle with the loading [31]. The inclined angle, $\varphi = \phi$ (**Figure 2b**), between the outward normal to the failure surface and the loading, can be determined by virtue of Mohr's theory as [9]

$$\phi = \frac{\pi}{4} + \frac{1}{2}\arcsin\frac{\sigma_{u,c}^{m} - \sigma_{u,t}^{m}}{2\sigma_{u,c}^{m}}. \tag{11}$$

The transverse tensile, transverse compressive, and transverse shear SCFs of the matrix in the composite are given as [9–11]

$$K_{22}^{t} = K_{22}(0), \tag{12.1}$$

$$K_{22}^{c} = K_{22}(\phi), \tag{12.2}$$

$$K_{23} = 2\sigma_{u,s}^{m}\sqrt{\frac{K_{22}^{t}K_{22}^{c}}{\sigma_{u,t}^{m}\sigma_{u,c}^{m}}}, \tag{12.3}$$

$\sigma_{u,t}^{m}$, $\sigma_{u,c}^{m}$, and $\sigma_{u,s}^{m}$ are the original tensile, compressive, and shear strengths of the matrix, respectively.

5.4 SCF under longitudinal shear

A longitudinal shear SCF of the matrix is given by [11]

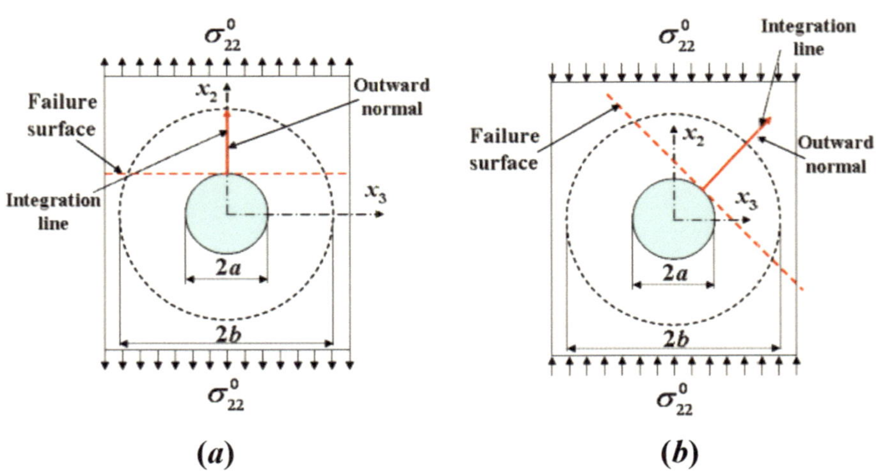

Figure 2.
Schematic of a RVE used in defining SCF of matrix in a composite subjected to (a) a transverse tension and (b) a transverse compression.

$$K_{12} = \left[1 - V_f \frac{G_{12}^f - G^m}{G_{12}^f + G^m} \left\{ \pi\sqrt{V_f} \left[\frac{1}{4V_f} - \frac{4}{128} - \frac{2}{512} V_f - \frac{5}{4096} V_f^{\,2} \right] - \frac{1}{3} \right\} \right] \frac{(V_f + a_{66} V_m)}{a_{66}}, \tag{13}$$

5.5 SCFs under equally biaxial transverse loads

Eq. (9) designates a general rule to determine any SCF of the matrix in the composite. Under an equally biaxial transverse tension or compression (**Figure 3**), a point-wise stress of the matrix in the x_2-direction is obtained through a coordinate transformation:

$$\tilde{\sigma}_{22}^m = \left[\tilde{\sigma}_{\rho\rho}^m(\sigma_{22}^0) + \tilde{\sigma}_{\rho\rho}^m(\sigma_{33}^0)\right]\cos^2\varphi + \left[\tilde{\sigma}_{\varphi\varphi}^m(\sigma_{22}^0) + \tilde{\sigma}_{\varphi\varphi}^m(\sigma_{33}^0)\right]\sin^2\varphi - \left[\tilde{\sigma}_{\rho\varphi}^m(\sigma_{22}^0) + \tilde{\sigma}_{\rho\varphi}^m(\sigma_{33}^0)\right]\sin 2\varphi, \tag{14}$$

where [41]

$$\tilde{\sigma}_{\rho\rho}^m(\sigma_{22}^0) = \frac{\sigma_{22}^0}{2}\left\{1 + Aa^2\rho^{-2} + \left[1 + B\left(4a^2\rho^{-2} - 3a^4\rho^{-4}\right)\right]\cos 2\varphi\right\}, \tag{15.1}$$

$$\tilde{\sigma}_{\varphi\varphi}^m(\sigma_{22}^0) = \frac{\sigma_{22}^0}{2}\left[1 - Aa^2\rho^{-2} - \left(1 - 3Ba^4\rho^{-4}\right)\cos 2\varphi\right], \tag{15.2}$$

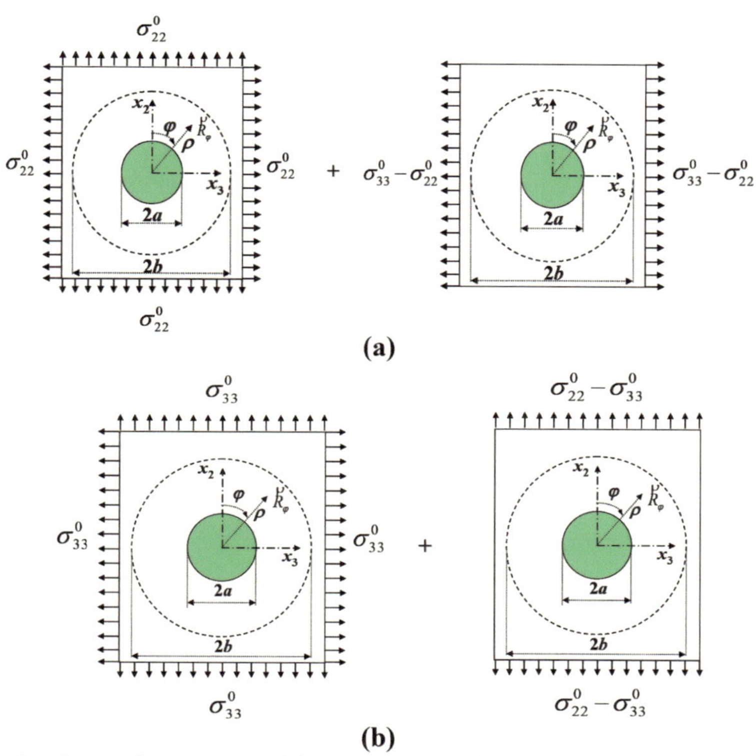

Figure 3.
Any biaxially transverse loads can be separated into superposition of an equally biaxial load and a uniaxial transverse tension either along x_3*-direction (a), if* $\sigma_{33}^0 - \sigma_{22}^0 > 0$*, or along* x_2*-direction (b), if* $\sigma_{22}^0 - \sigma_{33}^0 > 0$*.*

$$\tilde{\sigma}^m_{\rho\varphi}\left(\sigma^0_{22}\right) = -\frac{\sigma^0_{22}}{2}\left[1 - B\left(2a^2\rho^{-2} - 3a^4\rho^{-4}\right)\right]\sin 2\varphi. \tag{15.3}$$

The stresses $\tilde{\sigma}^m_{\rho\rho}\left(\sigma^0_{33}\right)$, $\tilde{\sigma}^m_{\varphi\varphi}\left(\sigma^0_{33}\right)$, and $\tilde{\sigma}^m_{\rho\varphi}\left(\sigma^0_{33}\right)$ are also given by Eqs. (15.1)–(15.3), respectively, as long as the σ^0_{22} in them is replaced by σ^0_{33} and φ by $\varphi = \varphi + \pi/2$. Substituting Eqs. (14) and (8.4) into Eq. (9), a biaxial transverse SCF of the matrix, $K^{Bi}_{22}(\varphi)$, is derived as

$$K^{Bi}_{22}(\varphi) = \frac{\sigma^0_{22}}{\left(\sigma^m_{22}\right)_{BM}} + \frac{A\sqrt{V_f}\left(\sigma^0_{22} + \sigma^0_{33}\right)}{2\left(\sigma^m_{22}\right)_{BM}}\cos 2\varphi + \frac{B\left(\sigma^0_{22} - \sigma^0_{33}\right)}{2\left(1 - \sqrt{V_f}\right)\left(\sigma^m_{22}\right)_{BM}}$$
$$\left[V_f^2\cos 4\varphi + 4V_f(\cos\varphi)^2(1 - 2\cos 2\varphi) + \sqrt{V_f}(2\cos\varphi + \cos 4\varphi)\right]$$

However, the failure surface orientation of a UD composite under an equally biaxial transverse tension or compression is indeterminate. For this reason, we can assume that the failure surface orientation under an equally biaxial transverse load is the same as that under a uniaxial transverse load. In other words, we have $\left(\sigma^0_{22} = \sigma^0_{33}\right)$.

$$K^{Bi,t}_{22} = K^{Bi,t}_{33} = \frac{(V_f + 0.3V_m)E^f_{22} + 0.7V_mE^m}{0.3E^f_{22} + 0.7E^m}\left(1 + A\sqrt{V_f}\right), \text{ if } \sigma^0_{33}>0, \tag{16.1}$$

$$K^{Bi,c}_{22} = K^{Bi,c}_{33} = \frac{(V_f + 0.3V_m)E^f_{22} + 0.7V_mE^m}{0.3E^f_{22} + 0.7E^m}\left(1 - A\sqrt{V_f\frac{\sigma^m_{u,c} - \sigma^m_{u,t}}{2\sigma^m_{u,c}}}\right), \text{ if } \sigma^0_{33} < 0. \tag{16.2}$$

5.6 SCFs subjected to any biaxial transverse loads

When the matrix is subjected to any biaxial transverse loads, we can always separate the loads into an equally biaxial transverse tension or compression plus a uniaxial transverse tension (**Figure 3**). The SCFs of the matrix are then determined accordingly.

5.7 Longitudinal normal SCF

No SCF exists in such a load case, since the resulting stresses in the matrix are uniform [7, 41].

6. Assessment on strength prediction

6.1 True stresses of the matrix

Let $\{\sigma^m_i\} = \left\{\sigma^m_{11}, \sigma^m_{22}, \sigma^m_{33}, \sigma^m_{23}, \sigma^m_{13}, \sigma^m_{12}\right\}^T$ be the homogenized stresses of the matrix in a UD composite calculated from a micromechanics model. The true stresses of the matrix, $\{\bar{\sigma}^m_i\} = \left\{\bar{\sigma}^m_{11}, \bar{\sigma}^m_{22}, \bar{\sigma}^m_{33}, \bar{\sigma}^m_{23}, \bar{\sigma}^m_{13}, \bar{\sigma}^m_{12}\right\}^T$, are determined as follows:

$$\{\overline{\sigma}_i^m\} = \{\sigma_{11}^m, K_{33}^{Bi}\sigma_{33}^m + K_{22}^t(\sigma_{22}^m - \sigma_{33}^m), K_{33}^{Bi}\sigma_{33}^m, K_{23}\sigma_{23}^m, K_{12}\sigma_{13}^m, K_{12}\sigma_{12}^m\}^T, \quad \text{if } \sigma_{22}^m \times \sigma_{33}^m \neq 0 \text{ and } (\sigma_{22}^m - \sigma_{33}^m) \geq 0, \tag{17.1}$$

$$\{\overline{\sigma}_i^m\} = \{\sigma_{11}^m, K_{22}^{Bi}\sigma_{22}^m, K_{22}^{Bi}\sigma_{22}^m + K_{22}^t(\sigma_{33}^m - \sigma_{22}^m), K_{23}\sigma_{23}^m, K_{12}\sigma_{13}^m, K_{12}\sigma_{12}^m\}^T, \quad \text{if } \sigma_{22}^m \times \sigma_{33}^m \neq 0 \text{ and } (\sigma_{33}^m - \sigma_{22}^m) \geq 0, \tag{17.2}$$

$$\{\overline{\sigma}_i^m\} = \{\sigma_{11}^m, K_{22}\sigma_{22}^m, 0, K_{23}\sigma_{23}^m, K_{12}\sigma_{13}^m, K_{12}\sigma_{12}^m\}^T, \text{ if } \sigma_{33}^m = 0, \tag{18.1}$$

$$\{\overline{\sigma}_i^m\} = \{\sigma_{11}^m, 0, K_{33}\sigma_{33}^m, K_{23}\sigma_{23}^m, K_{12}\sigma_{13}^m, K_{12}\sigma_{12}^m\}^T, \text{ if } \sigma_{22}^m = 0, \tag{18.2}$$

$$K_{22}^{Bi} = \begin{cases} K_{22}^{t,Bi}, \text{ if } \sigma_{22}^m > 0 \\ K_{22}^{c,Bi}, \text{ if } \sigma_{22}^m < 0 \end{cases}, \tag{19.1}$$

$$K_{33}^{Bi} = \begin{cases} K_{22}^{t,Bi}, \text{ if } \sigma_{33}^m > 0 \\ K_{22}^{c,Bi}, \text{ if } \sigma_{33}^m < 0 \end{cases}, \tag{19.2}$$

$$K_{22} = \begin{cases} K_{22}^t, \text{ if } \sigma_{22}^m > 0 \\ K_{22}^c, \text{ if } \sigma_{22}^m < 0 \end{cases}, \tag{19.3}$$

$$K_{33} = \begin{cases} K_{22}^t, \text{ if } \sigma_{33}^m > 0 \\ K_{22}^c, \text{ if } \sigma_{33}^m < 0 \end{cases}. \tag{19.4}$$

6.2 Uniaxial strength formulae

Bridging tensor elements of a micromechanics model for each of the nine UD composites can be calculated through Eq. (6), using the corresponding elastic moduli given in **Table B.3**.

Under a uniaxial load, only the internal stress component of a constituent (fiber or matrix) along the loading direction is dominant. The other stress components, if any, are negligibly small. This can be realized from the explicit Eqs. (8.1)–(8.6). Accordingly, a longitudinal failure of the composite is controlled mostly by a fiber failure, whereas all of the other failures are resulted from matrix failures. We only need to determine the following relationships:

$$\sigma_{11}^f = \lambda_1\sigma_{11}^0, \sigma_{22}^m = \lambda_2\sigma_{22}^0, \sigma_{23}^m = \lambda_3\sigma_{23}^0, \sigma_{12}^m = \lambda_4\sigma_{12}^0. \tag{20}$$

where λ_is are dependent on the bridging tensor and σ_{11}^0, σ_{22}^0, σ_{23}^0, and σ_{12}^0 are external loads applied individually to the composite once at a time. For each of the 9 composites, the λ_is calculated by the 12 models are summarized in **Table B.4**.

In terms of the data in **Table B.4**, the longitudinal tensile and compressive, transverse tensile and compressive, transverse shear, and longitudinal shear strengths of a UD composite are estimated through.

$$\sigma_{11}^{u,t} = \sigma_{u,t}^f/\lambda_1, \sigma_{11}^{u,c} = \sigma_{u,c}^f/\lambda_1, \sigma_{22}^{u,t} = \sigma_{u,t}^m/(K_{22}^t\lambda_2), \sigma_{22}^{u,c} = \sigma_{u,c}^m/(K_{22}^c\lambda_2), \quad \sigma_{23}^u = \sigma_{u,s}^m/(K_{23}\lambda_3), \text{ and } \sigma_{12}^u = \sigma_{u,s}^m/(K_{12}\lambda_4) \tag{21}$$

where K_{22}^t, K_{22}^c, K_{12}, and K_{23} are the transverse tensile, transverse compressive, longitudinal shear, and transverse shear SCFs of the matrix in the composite.

	E-glass LY556	E-glass MY750	AS4 3501–6	T300 BSL914C	IM7 8511–7	T300 PR319	AS epoxy	S2-glass epoxy	G400–800 5260
K_{12}	1.52	1.491	1.424	1.43	1.475	1.51	1.449	1.5	1.483
K_{23}	3.02	2.936	1.337	2.421	2.034	2.167	1.999	2.982	2.469
K_{22}^{t}	3.339	3.253	2.098	2.143	2.327	3.123	2.339	3.317	2.464
K_{22}^{c}	2.249	2.181	1.469	1.57	1.761	2.035	1.743	2.172	1.732
$K_{22}^{Bi,t}$	2.747	2.67	1.743	1.759	1.905	2.55	1.916	2.719	2.08
$K_{22}^{Bi,c}$	2.182	2.132	1.562	1.601	1.712	2.05	1.709	2.148	1.74

Table 2.
SCFs of the matrices in the nine composites.

6.3 Prediction assessment

Using the constituent data of **Table B.1**, all of the SCFs of the matrices in the nine composites are calculated as per Eqs. (12), (13), and (16). They are listed in **Table 2**. All of the SCFs only depend on the constituent properties and fiber volume fraction of a composite, since a perfect interface bonding has been implicitly assumed. **Table 2** shows that the transverse tensile SCF of the matrix is generally the biggest in a composite, whereas the transverse shear SCF is the second biggest or even bigger than all of the remaining SCFs in some composite. Further, the transverse tensile SCF can be greater than 3, implying that the classical SCF of a plate with a hole is not the upper limit for that of the matrix when the hole is filled with a fiber.

The predicted results are compared with the experimental measurements shown in **Table B.2**, and the averaged relative correlation errors for all of the 12 models are summarized in **Table 3**. It shows that Bridging Model is still overall the most accurate, although the accuracy difference between Bridging Model and the other

Model	N	Averaged error*	Error ratio	Rank	Model	N	Averaged error*	Error ratio	Rank
Bridging model	53	21.1%	1.0	**1**	Mori-Tanaka method	53	30.2%	1.43	**8**
Double inclusion (Digimat)	53	21.9%	1.04	**2**	Modified rule of mixture	53	30.7%	1.45	**10**
FE-square	53	23.1%	1.09	**3**	FE-square diagonal	53	31.9%	1.51	**11**
Chamis model	53	25.4%	1.20	**4**	FE-hexagonal	53	32%	1.52	**12**
FE-random	53	28.5%	1.30	**5**	Self-consistent method	53	32.7%	1.54	**13**
Hill-Hashin-C-L model	18	30.1%	1.43	**6**	Rule of mixture method	53	44.5%	2.11	**14**
Halpin-Tsai formulae	53	30.1%	1.43	**6**	Eshelby's method	53	45.1%	2.14	**15**
Generalized self-consistent	53	30.2%	1.43	**8**					

$^{*}=\frac{1}{N}\sum_{i=1}^{N} abs(error)_i$.

Table 3.
Overall averaged errors in prediction of the uniaxial strengths of the nine UD composites by different models.

top three models is insignificant. Compared **Table 3** with **Table 1**, the ranking order of the top four theories for both the stiffness and strength predictions is essentially the same, with only a minor difference in the ranking order from stiffness and strength predictions by the FE-square and double inclusion method.

The largest correlation error in the strength prediction is still assumed by Eshelby's method, which is 45.1%. Another model gaining a correlation error of more than 40% is rule of mixture method. All of the theories under consideration for the strength predictions can be classified into three classes, according to their accuracies attained. The first class exhibits the highest accuracy. It consists of four methods, which are Bridging Model, double inclusion method, the FE-square, and Chamis model, with a correlation error in between 21.1% and 25.4%. The second class is moderate in accuracy performance. Most of the models, i.e., the FE-random, Hill-Hashin-Christensen-Lo model, Halpin-Tsai formulae, generalized self-consistent method, Mori-Tanaka method, modified rule of mixture method, the FE-square diagonal, the FE-hexagonal and self-consistent method, are within this class. Their correlation errors vary from 27.4% to 32.7%. The third class possesses the lowest prediction accuracy, consisting of two models, i.e., rule of mixture method and Eshelby's method. Looking back at **Table 1**, the classification of the three classes of the micromechanics models for the stiffness predictions is also valid.

If no SCFs of the matrix are taken into account, i.e., if $K_{22}^{t} = K_{22}^{c} = K_{12} = K_{23} \equiv 1$ are assumed in Eq. (21), the overall correlation error by a model from the first or the second class is much greater. Consider, e.g., Bridging Model. Without the SCFs, the correlation error between the predicted and measured transverse tensile, transverse compressive, transverse shear, and longitudinal shear strengths of the 9 composites is 115.3%, 5.22 times greater than that when the SCFs are taken into account. It is noted that the longitudinal strength predictions have been excluded in this latter comparison. Hence, the most critical factor to influence the overall strength prediction is the SCFs of the matrix in the composite.

7. Additional comments

7.1 Consistency

Eqs. (1)–(6) are valid for both 2D and 3D stress states. Any micromechanics model can result in two sets of internal stress formulae, i.e., 2D and 3D formulae, respectively. Let the composite be subjected to a planar stress state $\{\sigma_{11}^{0}, \sigma_{22}^{0}, \sigma_{12}^{0}\}$. The resulting internal stresses in the fiber and matrix by the 2D formulae are represented as $\{\sigma_{11}^{f,2D}, \sigma_{22}^{f,2D}, \sigma_{33}^{f,2D}, \sigma_{23}^{f,2D}, \sigma_{13}^{f,2D}, \sigma_{12}^{f,2D}\}$ and $\{\sigma_{11}^{m,2D}, \sigma_{22}^{m,2D}, \sigma_{33}^{m,2D}, \sigma_{23}^{m,2D}, \sigma_{13}^{m,2D}, \sigma_{12}^{m,2D}\}$, where $\sigma_{33}^{f,2D} = \sigma_{23}^{f,2D} = \sigma_{13}^{f,2D} = \sigma_{33}^{m,2D} = \sigma_{23}^{m,2D} = \sigma_{13}^{m,2D} = 0$. On the other hand, using the 3D stress formulae and applying a load combination $\{\sigma_{11}^{0}, \sigma_{22}^{0}, 0, 0, 0, \sigma_{12}^{0}\}$, the internal stresses thus obtained are denoted as $\{\sigma_{11}^{f,3D}, \sigma_{22}^{f,3D}, \sigma_{33}^{f,3D}, \sigma_{23}^{f,3D}, \sigma_{13}^{f,3D}, \sigma_{12}^{f,3D}\}$ and $\{\sigma_{11}^{m,3D}, \sigma_{22}^{m,3D}, \sigma_{33}^{m,3D}, \sigma_{23}^{m,3D}, \sigma_{13}^{m,3D}, \sigma_{12}^{m,3D}\}$. If $\sigma_{ij}^{f,2D} = \sigma_{ij}^{f,3D}$ and $\sigma_{ij}^{m,2D} = \sigma_{ij}^{m,3D}$ for all i and j, the corresponding micromechanics model is said to be consistent in the internal stress calculation.

A necessary and sufficient condition for a micromechanics model to be consistent is that its bridging tensor is always in an upper triangular form. If, e.g., $A_{32} \neq 0$, we will get from Eqs. (3) and (4) that $\sigma_{33}^{f} = B_{32}\sigma_{22}^{0} \neq 0$ and $\sigma_{33}^{m} = A_{32}\sigma_{22}^{f} + A_{33}\sigma_{33}^{f} \neq 0$, where $[B_{ij}] = (V_f[I] + V_m[A_{ij}])^{-1}$. The bridging tensor of Bridging Model, by definition, is always upper triangular, even when a constituent, e.g., matrix undergoes a plastic deformation [16]. On the other hand, the bridging tensors of all of the other models for the nine composites are not upper triangular. Hence, all of the other models are not consistent. The non-consistency implies that the homogenized internal stresses should be calculated using the full 3D stress formulae, even though the composite is subjected to a uniaxial load. To apply an analytical model, other than Bridging Model, the 3D compliance tensors of the fiber, matrix, and the composite should be used in Eq. (6) to obtain the 3D bridging tensor. If a numerical method is applied to predict a composite property, a 3D rather than 2D RVE geometry should be discretized.

7.2 Accuracy estimation

It is known that the elasticity of UD composites is essentially matured. This means that the accuracy in both experimental measurement and micromechanics prediction for the elastic properties of a UD composite is likely not improvable significantly, unless a revolutionary change in the processing technology for a composite occurs. Considering the measurement deviations for the fiber, matrix, and composite properties as well as for the fiber volume fraction and in light of **Table 1**, it can draw a conclusion that an overall correlation error of 10% would be the one attainable in the composite stiffness prediction if only the original constituent information is available. As more than double of the constituent data together with more other information are required in a strength prediction, a reasonable correlation error in this latter case that can be expected should be more than 10% and mostly up to 20%.

The individual correlation error for each of the uniaxial strengths of the nine UD composites by Bridging Model is calculated and is shown in **Table 4**. Evidently, the predictions by the current theory for the longitudinal tensile, longitudinal shear, and transverse shear strengths of the composites are good enough, whereas those for the other three strength properties are either bad or not very satisfactory. Improvement in the prediction accuracy for the latter three uniaxial strengths is expected.

7.3 Improvement in strength prediction

From **Table 4**, the largest correlation error comes from the prediction of a transverse tensile strength. This is attributed to a crack occurred in between the fiber and matrix interface. There must be some composites in which the fiber and matrix interfaces were already debonded before an ultimate failure under a transverse tensile load. Many researches in the literature have confirmed that an interface debonding has the greatest influence on the transverse tensile strength of a composite [42–44]. Therefore, a true stress component of the matrix corresponding to a transverse tension must take an interface debonding into account [11].

The second largest error is in the prediction of a longitudinal compressive strength. Only a strength failure has been considered in this work for a composite

Longitudinal tensile strength	Longitudinal compressive strength	Longitudinal shear strength	Transverse tensile strength	Transverse compressive strength	Transverse shear strength	Overall
11%	25.1%	13.1%	39.2%	23.2%	14.3%	21.1%

Table 4.
Averaged correlation errors of Bridging Model for the individul uniaxial strengths of the nine UD composites.

subjected to a longitudinal compression. Existing evidences show that a longitudinal compressive failure is frequently caused by a kink-band or microbuckling [45–47], due to an initial fiber misalignment. A micromechanics approach for a kink-band failure only using the original fiber and matrix properties together with the initial fiber misalignment angle has been achieved very recently [48]. However, a fiber misalignment angle is in most cases an in situ parameter and is difficult to be accurately measured. A more suitable way is to retrieve it from a measured longitudinal compressive strength parameter of the composite. On the other hand, this parameter can also be used to adjust the fiber compressive strength to improve the correlation accuracy.

The third correlation error, which is greater than 20%, occurs in the prediction for the transverse compressive strengths of the composites. Most probably, this error is attributed mainly to an inaccurate measurement/determination of a matrix compressive strength. It is known that among the three uniaxial strength parameters of a matrix especially a ductile polymer or metal matrix material, the compressive strength is the most difficult to be measured. Sometimes, one even cannot obtain a fracture load when a cylinder sample is compressively tested. Further study is needed to determine a matrix compressive strength.

An interesting phenomenon behind **Table 4** is that either longitudinal or transverse shear strength can be sufficiently well predicted based on a perfect interface bonding assumption. Undoubtedly, an interface debonding may occur when the composite is subjected to a shear load. But the interface debonding has insignificant effect on the shear as well as on any other kind of load carrying ability of a composite except for the transverse tension, as seen in the subsequent section.

8. Failure prediction with a debonded interface

8.1 Transverse tensile SCF of the matrix after interface crack

All of the SCFs presented in the preceding section are based on an assumption that the fiber and matrix interface has a perfect bonding up to a composite failure. In other words, the point-wise displacements and the point-wise stresses of the fiber and matrix on their common boundary are continuous. In most cases, an interface debonding or crack can occur before an ultimate failure of the composite. However, **Table 4** suggests that only the transverse tensile load sustaining ability of the composite is influenced heavily by the interface crack or debonding. The transverse tensile SCF of the matrix after the interface crack (**Figure 4**) must be taken into account in a failure prediction, in general. This SCF has been derived in Ref. [11], which is summarized below.

$$\hat{K}_{22}^t = \hat{K}_{22}^t(\psi) = \mathrm{Re}\left\{e^{-2i\psi}M\left(be^{i\psi}\right)\left(a^2/b - b\right) - e^{-i\psi}\left(N_2 - N_1\left(\frac{a^2}{b}e^{-i\psi}\right)\right)\right.$$
$$\left. + e^{-i\psi}\left(2 + e^{-2i\psi}\right)\left[N\left(be^{i\psi}\right) - N_3\right]\right\}\frac{(V_f + 0.3V_m)E_{22}^f + 0.7V_mE^m}{2(b-a)\left(0.3E_{22}^f + 0.7E^m\right)}, \tag{22.1}$$

$$N(z) = Fz + \frac{a^2k}{z} - \left(z - ae^{i\psi}\right)^{0.5+i\lambda}\left(z - ae^{-i\psi}\right)^{0.5-i\lambda}\left[(F - 0.5) - \frac{D}{a^2z}\right], \tag{22.2}$$

$$N_1(z) = Fz + \frac{a^2k}{z} + \frac{1}{\xi}\left(z - ae^{i\psi}\right)^{0.5+i\lambda}\left(z - ae^{-i\psi}\right)^{0.5-i\lambda}\left[(F - 0.5) - \frac{D}{a^2z}\right], \tag{22.3}$$

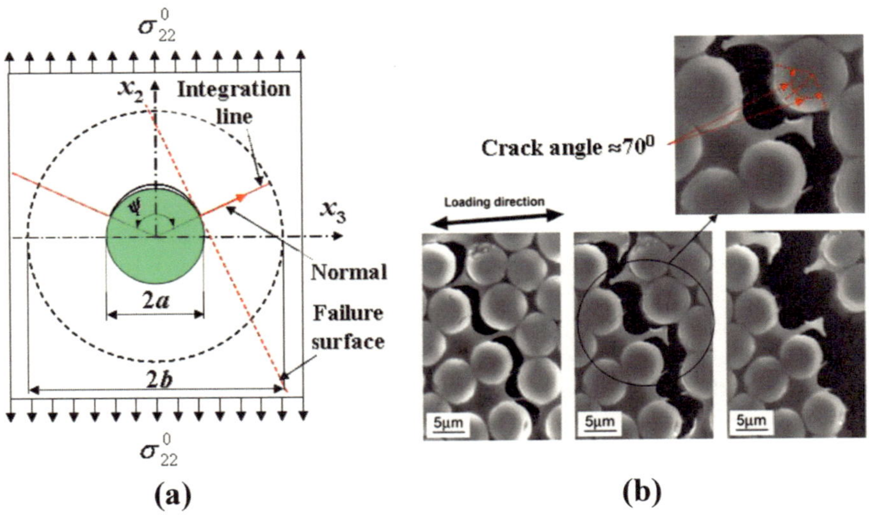

Figure 4.
(a) Schematic failure of a transverse tensile-loaded composite with an interface crack, (b) failure locus of a composite after interface cracks [42].

$$N_2 = aFe^{-i\psi} + ake^{i\psi}, N_3 = Fae^{i\psi} + e^{-i\psi}ak, \tag{22.4}$$

$$M(z) = F - \frac{a^2k}{z^2} - \left[(F - 0.5)z + H + \frac{C}{z} + \frac{D}{z^2}\right]\chi(z), \tag{22.5}$$

$$F = \frac{1 - (\cos\psi + 2\lambda\sin\psi)\exp\left[2\lambda(\pi - \psi)\right] + (1 - k)(1 + 4\lambda^2)\sin^2\psi}{\frac{4}{k} - 2 - 2(\cos\psi + 2\lambda\sin\psi)\exp\left[2\lambda(\pi - \psi)\right]}, \tag{22.6}$$

$$H = a(\cos\psi + 2\lambda\sin\psi)(0.5 - F), \tag{22.7}$$

$$C = (k - 1)(\cos\psi - 2\lambda\sin\psi)a^2\exp\left[2\lambda(\psi - \pi)\right], \tag{22.8}$$

$$D = (1 - k)a^3\exp\left[2\lambda(\psi - \pi)\right], \tag{22.9}$$

$$\chi(z) = \left(z - ae^{i\psi}\right)^{-0.5+i\lambda}\left(z - ae^{-i\psi}\right)^{-0.5-i\lambda}, \tag{22.10}$$

$$k = \frac{G^m(1 + \kappa_2)}{(1 + \xi)\left(G^m + \kappa_1 G_{23}^f\right)}, \lambda = -(ln\xi)/(2\pi), \xi = \left(G_{23}^f + \kappa_2 G^m\right)/\left(G^m + \kappa_1 G_{23}^f\right),$$

$$\kappa_1 = 3 - 4v^m, \kappa_2 = \frac{3 - v_{23}^f - 4v_{12}^f v_{21}^f}{1 + v_{23}^f}, \tag{22.11}$$

$$b = a\Big/\sqrt{V_f}. \tag{22.12}$$

In the above, ψ is the half of the crack angle, which is determined from

$$\mathrm{Re}\left\{\left(G_0 - \frac{1}{k} - \frac{2(1 - k)}{k\exp(i\varphi)}\exp\left[2\lambda(\psi - \pi)\right]\right)R\left(e^{i\varphi}\right)\right\}_{\varphi = \psi - \gamma} = 0, \tag{23.1}$$

$$R(\exp(i\varphi)) = \left[\exp(i(\varphi)) - e^{i\psi}\right]^{0.5+i\lambda}\left[\exp(i(\varphi)) - e^{-i\psi}\right]^{0.5-i\lambda}\exp(-i(\varphi)), \tag{23.2}$$

$$G_0 = \frac{1 - (\cos\psi + 2\lambda \sin\psi)\exp[2\lambda(\pi - \psi)] + (1-k)(1+4\lambda^2)\sin^2\psi}{2 - k - k(\cos\psi + 2\lambda\sin\psi)\exp[2\lambda(\pi-\psi)]}, \tag{23.3}$$

$$\gamma = \begin{cases} \dfrac{2\lambda(J_1^2 + J_2^2)}{J_1^2 + J_2^2 - 2J_2J_3}, & if\, \xi < 1 \\ -\dfrac{2\lambda(J_1^2 + J_2^2)}{J_1^2 + J_2^2 - 2J_2J_3}, & if\, \xi > 1 \end{cases}, \tag{23.4}$$

$$J_1 = kG_0 - 1 - 2(1-k)\xi\exp(2\lambda\psi)\cos(\psi), \tag{23.5}$$

$$J_2 = 2(1-k)\xi\exp(2\lambda\psi)\sin(\psi), \tag{23.6}$$

$$J_3 = 2(1-k)\xi\exp(2\lambda\psi)[J_1\cos(\psi) - J_2\sin(\psi)]/J_2. \tag{23.7}$$

If $\xi = 1$, no solution for ψ is obtainable from Eq. (23). The corresponding interface crack is called a singular crack. But one can always adjust the fiber or the matrix properties involved so that $\xi \neq 1$, since experimental deviations exist in measurement of them.

8.2 Interface crack detection

Let a UD composite be subjected to a transverse tension, σ_{22}^0, up to an ultimate failure. The measured transverse tensile strength of the composite is Y. Suppose that the fiber/matrix interface of the composite is initially bonded perfectly. When the load is increased to a critical level, e.g., $\hat{\sigma}_{22}^0$, a stable crack with a central angle of 2ψ occurs on the interface. Many reports have pointed out that an unstable propagation from an initial interface crack to the last stable angle is short [42, 49, 50], with no significant change in the applied load. Thus, we can safely assume that at a transverse load level smaller than $\hat{\sigma}_{22}^0$ the interface is in perfect bonding.

From Eq. (8.4), the transverse stress in the matrix when the crack occurs reads

$$\hat{\sigma}_{22}^m = \frac{0.3E_{22}^f + 0.7E^m}{(V_f + 0.3V_m)E_{22}^f + 0.7V_mE^m}\hat{\sigma}_{22}^0. \tag{24.1}$$

Further, the longitudinal stress of the matrix at the critical load level is obtained from Eq. (8.2) as

$$\hat{\sigma}_{11}^m = \frac{V_f a_{12}}{(V_f + V_m a_{11})(V_f + V_m a_{22})}\hat{\sigma}_{22}^0. \tag{24.2}$$

No other stress in the matrix exists. Supposing that the transverse matrix stress corresponding to the composite failure is denoted by $\overline{\overline{\sigma}}_{22}^m$, one has.

$$\hat{K}_{22}^t\left(\overline{\overline{\sigma}}_{22}^m - \hat{\sigma}_{22}^m\right) + K_{22}^t\hat{\sigma}_{22}^m = \sigma_{u,t}^m, \tag{25.1}$$

where

$$\overline{\overline{\sigma}}_{22}^m - \hat{\sigma}_{22}^m = \frac{0.3E_{22}^f + 0.7E^m}{(V_f + 0.3V_m)E_{22}^f + 0.7V_mE^m}\left(Y - \hat{\sigma}_{22}^0\right). \tag{25.2}$$

From Eqs. (24.1), (25.1), and (25.2), the critical transverse tensile load is found to be

$$\hat{\sigma}_{22}^{0} \frac{= \hat{K}_{22}^{t} Y^{t}}{\hat{K}_{22}} - K_{22}^{t} - \frac{(V_f + 0.3V_m)E_{22}^{f} + 0.7V_m E^m}{\left(0.3E_{22}^{f} + 0.7E^m\right)\left(\hat{K}_{22}^{t} - K_{22}^{t}\right)} \sigma_{u,t}^{m}. \tag{26}$$

If it is equal to or greater than the transverse tensile strength, Y, the fiber and matrix system is said to have a perfect interface bonding up to failure. Otherwise, the system will undergo an earlier interface crack and a further interface modification is preferred.

Under any arbitrary load condition, an interface crack occurs in the composite if and only if

$$\left(\bar{\sigma}_m^1\right)_l > 0 \tag{27.1}$$

and

$$\left(\bar{\sigma}_e^m\right)_l \geq \hat{\sigma}_e^m, \tag{27.2}$$

where

$$\hat{\sigma}_e^m = \sqrt{\left(\hat{\sigma}_{11}^m\right)^2 + \left(K_{22}^t \hat{\sigma}_{22}^m\right)^2 - K_{22}^t \hat{\sigma}_{11}^m \hat{\sigma}_{22}^m} \tag{28}$$

is the critical Mises stress. $\left(\bar{\sigma}_e^m\right)_l$ and $\left(\bar{\sigma}_m^1\right)_l$ are the Mises and the first principal stresses of the matrix obtained from the current load-step true stresses. For instance, when a planar load is applied to the composite, the current Mises true stress is given by

$$\left(\bar{\sigma}_e^m\right)_l = \sqrt{\left(\bar{\sigma}_{11}^m\right)_l^2 + \left(\bar{\sigma}_{22}^m\right)_l^2 - \left(\bar{\sigma}_{11}^m\right)_l\left(\bar{\sigma}_{22}^m\right)_l + 3\left(\bar{\sigma}_{12}^m\right)_l^2}, \tag{29}$$

$$\left(\bar{\sigma}_{11}^m\right)_l = \left(\bar{\sigma}_{11}^m\right)_{l-1} + d\sigma_{11}^m, \tag{30.1}$$

$$\left(\bar{\sigma}_{22}^m\right)_l = \left(\bar{\sigma}_{22}^m\right)_{l-1} + K_{22}^{eq} d\sigma_{22}^m, \tag{30.2}$$

$$\left(\bar{\sigma}_{12}^m\right)_l = \left(\bar{\sigma}_{12}^m\right)_{l-1} + K_{12} d\sigma_{12}^m, \tag{30.3}$$

$$K_{22}^{eq} = \begin{cases} K_{22}^t, \text{ if } d\sigma_{22}^m > 0 \text{ and } \left(\bar{\sigma}_e^m\right)_{l-1} < \hat{\sigma}_e^m \\ \hat{K}_{22}^t, \text{ if } d\sigma_{22}^m > 0 \text{ and } \left(\bar{\sigma}_e^m\right)_{l-1} \geq \hat{\sigma}_e^m \\ K_{22}^c, \text{ if } d\sigma_{22}^m < 0 \end{cases}. \tag{31}$$

The homogeneous stress increments, $\{d\sigma_{11}^m, d\sigma_{22}^m, d\sigma_{12}^m\}$, are calculated from Eqs. (8.2) and (8.4) in which $\{\sigma_{11}^0, \sigma_{22}^0, \sigma_{12}^0\}$ are replaced by $\{d\sigma_{11}^0, d\sigma_{22}^0, d\sigma_{12}^0\}$.

Using the data of **Tables B.1** and **B.2**, the transverse tensile SCFs of the nine UD composites after the interface crack together with the crack (half) angles are

	E-glass LY556	E-glass MY750	AS4 3501–6	T300 BSL914C	IM7 8551–7	T300 PR319	AS epoxy	S2-glass epoxy	G40–800 5260
$\hat{K}_{22}^t$	7.69	7.22	4.95	5.04	5.41	6.97	5.43	7.34	5.68
ψ	71.8°	71.9°	73.9°	73.9°	73.4°	72°	73.3°	71.8°	72.8°
$\hat{\sigma}_{22}^0$ (MPa)	20.3	28.2	44.4	4.7	72.5	33.5	19	74.3	93.2
$\hat{\sigma}_e^m$ (MPa)	28.6	39.8	53.9	5.76	90.6	46.3	24	105.1	119.4

Table 5.
Transverse tensile SCFs and other relevant parameter of the nine composites after interface debonding.

calculated from Eqs. (22) and (23). The critical transverse and Mises stresses are also obtained from Eqs. (26) and (28), respectively. They are summarized in **Table 5**. It is seen that the half crack angles of the carbon and glass fiber matrix interfaces under a transverse tension are close to 70°, consistent with the measured result shown in **Figure 4b**. Comparing the resulting $\hat{\sigma}_{22}^{0}$ with the corresponding measured transverse tensile strengths, one can see that four composites, i.e., AS4/3501-6, IM7/8551-7, S2-Glass/Epoxy, and G40-800/5260 systems, have or nearly have a perfect interface bonding up to failure. The remaining five composites will undergo an interface debonding early before failure. There are two composite systems, S2-Glass/Epoxy and G40-800/5260, having a more than enough interface bonding strength, implying that more than enough efforts might have been paid.

8.3 Off-axial strength prediction

A composite strength is assumed if either a fiber or a matrix failure is attained. A matrix failure is detected through, e.g., Tsai-Wu's criterion (as isotropic materials are a subset of anisotropic composites), whereas a fiber failure is assessed by the generalized maximum normal stress failure criteria [16], through the following expressions:

$$F_1\left[\left(\bar{\sigma}_{11}^m\right)_l^2+\left(\bar{\sigma}_{22}^m\right)_l^2-\left(\bar{\sigma}_{11}^m\right)_l\left(\bar{\sigma}_{22}^m\right)_l\right]+F_2\left(\bar{\sigma}_{12}^m\right)_l^2+F_3\left[\left(\bar{\sigma}_{11}^m\right)_l+\left(\bar{\sigma}_{22}^m\right)_l\right]\geq 1, \tag{32.1}$$

$$F_1=1/\left(\sigma_{u,t}^m\sigma_{u,c}^m\right), F_2=1/\left(\sigma_{u,s}^m\right)^2, F_3=1/\sigma_{u,t}^m-1/\sigma_{u,c}^m. \tag{32.2}$$

$$\left(\sigma_{eq,t}^f\right)_l\geq\sigma_{u,t}^f \text{ or } \left(\sigma_{eq,c}^f\right)_l\geq-\sigma_{u,c}^f, \tag{33.1}$$

$$\left(\sigma_{eq,t}^f\right)_l=\begin{cases}\left(\sigma_f^1\right)_l, & if\ \left(\sigma_f^3\right)_l<0,\\ \left[\left(\sigma_f^1\right)_l^3+\left(\sigma_f^2\right)_l^3\right]^{\frac{1}{3}}, & if\ \left(\sigma_f^3\right)_l=0\end{cases}, \tag{33.2}$$

$$\left(\sigma_{eq,c}^f\right)_l=\begin{cases}\left(\sigma_f^3\right)_l, & if\ \left(\sigma_f^1\right)_l>0,\\ \left(\sigma_f^3\right)_l-\left(\sigma_f^1\right)_l & if\ \left(\sigma_f^1\right)_l\leq 0\end{cases}. \tag{33.3}$$

$\left(\sigma_f^1\right)_l$, $\left(\sigma_f^2\right)_l$, and $\left(\sigma_f^3\right)_l$ $(\sigma_f^1\geq\sigma_f^2\geq\sigma_f^1)$ are the three principal stresses of the fiber calculated from

$$\left(\sigma_{11}^f\right)_l=\left(\sigma_{11}^f\right)_{l-1}+d\sigma_{11}^f, \tag{34.1}$$

$$\left(\sigma_{22}^f\right)_l=\left(\sigma_{22}^f\right)_{l-1}+d\sigma_{22}^f, \tag{34.2}$$

$$\left(\sigma_{12}^f\right)_l=\left(\sigma_{12}^f\right)_{l-1}+d\sigma_{12}^f. \tag{34.3}$$

$\sigma_{u,t}^f$ and $\sigma_{u,c}^f$ are longitudinal tensile and compressive strengths of the fiber, respectively.

Two UD composites, Kevlar-49/epoxy and E-glass/8804 epoxy systems, were subjected to off-axial tensile load up to failure. Constituent properties and transverse tensile strengths of the two composites as well as fiber volume fractions were provided in Ref. [51, 52] and cited in **Table 6**. From them, the SCFs of the matrices and the critical Mises stresses can be calculated and are also shown in the table. The predicted off-axial strengths of the Kevlar-49/epoxy and E-glass/8804 composites

	E_{11}^f (GPa)	E_{22}^f (GPa)	G_{12}^f (GPa)	ν_{12}^f	ν_{23}^f	$\sigma_{u,t}^f$ (GPa)	E^m (GPa)	ν^m	$\sigma_{u,t}^m$ (MPa)	$\sigma_{u,c}^m$ (MPa)	$\sigma_{u,s}^m$ (MPa)	V_f	K_{22}^t	K_{22}^c	$\hat{K}_{22}^t$	K_{12}	$\hat{\sigma}_e^m$ (MPa)
Kevlar-49/ epoxy	124.1	4.1	2.9	0.35	0.35	2.06	3.45	0.35	69	120	50	0.55	1.08	1.07	2.74	1.17	1.1
E-glass/ 8804	71	71	28.2	0.26	0.26	1.5	3.1	0.29	70	86	39	0.51	2.97	2.02	5.6	1.38	59.6

Table 6.
Constituent properties and resulting matrix parameters of two UD composites [51, 52].

are plotted in **Figures 5** and **6**, respectively. The SCFs in the Kevlar fiber system with a perfect interface bonding are close to 1, because the transverse modulus of the Kevlar fiber is comparable to that of the matrix. Nevertheless, the transverse tensile SCF of the matrix in the Kevlar fiber system after the interface crack is still significantly higher than that with the perfect interface bonding. Both of the critical transverse loads of the composites, $\hat{\sigma}_{22}^{0}$ = 1.2 MPa for the Kevlar and $\hat{\sigma}_{22}^{0}$ = 42.4 MPa for the glass composites, are smaller than the corresponding transverse tensile strengths, 27.7 and 45.3 MPa, respectively, and the two composites will undergo an interface crack. However, the Kevlar fiber system will crack much earlier than the E-glass fiber system does when subjected to a transverse tensile load.

Given an off-axial tensile load increment, $d\sigma_\theta$, where θ is the off-axial angle, the stress increments $\{d\sigma_{11}^{0}, d\sigma_{22}^{0}, d\sigma_{12}^{0}\}$ are obtained through a coordinate transformation, and the interface cracking load, $\hat{\sigma}_{\theta}^{0}$, can be determined through Eq. (27). Consider the Kevlar fiber composite, for instance, and take θ = 30°. The homogenized stresses of the matrix are obtained from Eqs. (8.2), (8.4), and (8.6) as $\sigma_{11}^{m} = 0.115\sigma_\theta$, $\sigma_{22}^{m} = 0.234\sigma_\theta$, and $\sigma_{12}^{m} = -0.32\sigma_\theta$, whereas the true stresses before the interface crack are $\bar{\sigma}_{11}^{m}$ = $0.115\sigma_\theta$, $\bar{\sigma}_{22}^{m} = 0.253\sigma_\theta$, and $\bar{\sigma}_{12}^{m} = -0.374\sigma_\theta$. Hence, the interface cracking load $\hat{\sigma}_{30^0}^{0}$ = 1.11/0.684 = 1.623 MPa.

After the interface crack, only the transverse tensile stress increment of the matrix will be amplified with a different SCF. The true stresses of the matrix in the Kevlar fiber composite with θ = 30° and after the interface crack are given by $\bar{\sigma}_{11}^{m}$ = $0.115\sigma_\theta$, $\bar{\sigma}_{22}^{m}$= 0.41 + $0.641(\sigma_\theta - 1.623)$, and $\bar{\sigma}_{12}^{m} = -0.374\sigma_\theta$. As the matrix failure occurs first, the predicted ultimate off-axial tensile strength from Eq. (32) is $\sigma_{30^0}^{u,t}$ = 80.5 MPa, very close to the measured one, 83.4 MPa [53].

The measured data for the Kevlar and the glass composites taken from Pindera et al. [53] and Mayes et al. [52] are also shown on **Figures 5** and **6**, respectively. In order to display the predicted results at most off-axial angles more clearly, the predictions at angles smaller than 10° are not included in the figures. Three kinds of predictions have been made. One is with a perfect interface bonding assumption, another without any SCF of the matrix considered, and the third is incorporated with an interface crack. As expected, the predictions without any SCF are far away from the experiments at most off-axial angles, whereas those with the interface

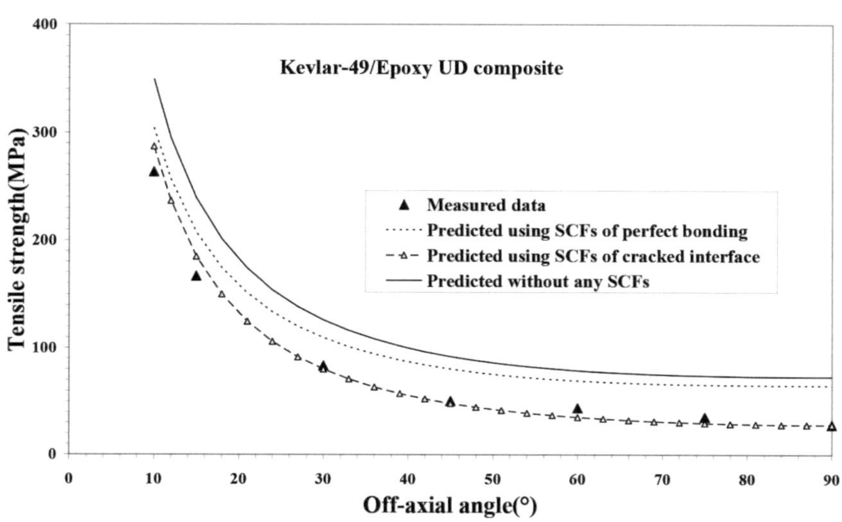

Figure 5.
Comparison of different schemes' predictions with experiments [53] for off-axial tensile strengths of a Kevlar-49/epoxy UD composite.

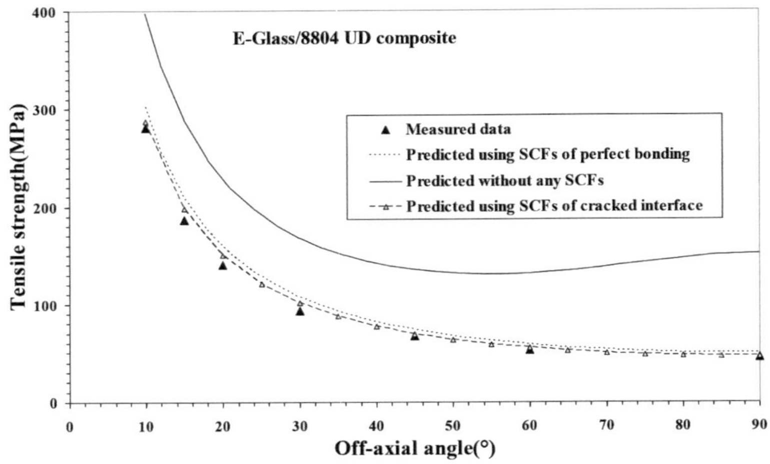

Figure 6.
Comparison of different schemes' predictions with experiments [53] for off-axial tensile strengths of an E-glass/ 8804 UD composite.

crack incorporated agree the best with the measured data. The perfect bonding assumption for both of the composites results in the predictions lied in between the other two kinds of predictions. Whereas the perfect bonding assumption up to a composite failure for the E-glass fiber system is good enough (**Figure 6**), the same assumption for the Kevlar fiber system generates significant prediction errors in general (**Figure 5**). This is because the E-glass fiber system under consideration has a critical transverse load (42.4 MPa) quite close to the transverse tensile strength, 45.3 MPa. On the other hand, the Kevlar fiber system can only sustain a transverse tensile load up to 1.2 MPa before an interface crack, which is very small compared to the transverse tensile strength, 27.7 MPa. This is consistent with a common observation that a Kevlar fiber-reinforced polymer matrix composite generally undergoes a much earlier interface debonding before an ultimate failure.

It is noticed that the three kinds of predictions arrive at the same longitudinal strength for each composite, i.e., 1137 MPa for the Kevlar fiber and 852 MPa for the E-glass fiber composites. Both of them correlate well with the experimental data, i.e., 1142 MPa for the Kevlar composite [53] and 817.5 MPa for the glass composite [52]. However, both of them have already undergone an interface crack ($\hat{\sigma}^{0}_{0^0}$ = 22.3 MPa for the Kevlar and 725.4 MPa for the glass composites) before the longitudinal strength is attained. Both of the examples confirm that only a transverse tensile load carrying capacity is influenced by an interface crack.

9. Conclusion

Micromechanical failure and strength prediction of a UD composite is systematically described in this chapter. The internal stresses in the fiber and matrix must be calculated at first. Although various methods are available for this purpose, a thorough compassion study has shown that Bridging Model is the most potential owing to its simplicity with closed-form formulae, high accuracy, and the unique feature of consistency in the internal stress calculation. These stresses are homogenized quantities. They must be converted into true values before being used for a failure assessment against the original strength data of the fiber and matrix. Otherwise, a predicted strength of the composite may be far away from the experiment. The difference in the overall prediction accuracies between use of the homogenized and the true stresses can be as large as more than five times. As the homogenized and the true stresses in the fiber are the same, the true stress determination for the matrix becomes essential. It has been demonstrated in the chapter that the true stresses of the matrix under any load condition are obtained by multiplying its homogenized counterparts with SCFs of the matrix in the composites. Such an SCF cannot be determined following a classical way. Instead, it must be defined as a line-averaged stress divided by a volume-averaged quantity. All of the SCFs subjected to various kinds of loading and with a perfect interface bonding have been presented in the chapter. It is known that an interface debonding only has a significant influence on the transverse tensile load carrying ability, and the transverse tensile SCF of the matrix after the interface debonding has also been obtained. The theory is equally well applicable to the failure and strength prediction of any other continuous fiber-reinforced composite, once it is subdivided into a combination of UD composites.

Acknowledgements

Partial financial supports from the National Natural Science Foundation of China with grant numbers of 11832014 and 11472192 are greatly acknowledged.

Appendix A. Formulae of micromechanics models

A.1 Eshelby's method

The stiffness tensor of the composite by this method is given by [15]

$$[K_{ij}] = [S_{ij}]^{-1} = \left[K_{ij}^m\right] + V_f\left(\left[K_{ij}^f\right] - \left[K_{ij}^m\right]\right)\left\{[I] + [L_{ij}]\left[S_{ij}^m\right]\left(\left[K_{ij}^f\right] - \left[K_{ij}^m\right]\right)\right\}^{-1}, \quad \text{(A.1)}$$

$\left[K_{ij}^f\right]$ and $\left[K_{ij}^m\right]$ are the stiffness tensors of the fiber and matrix, respectively, and $[L_{ij}]$ is the Eshelby's tensor reading [16]

$$[L_{ij}] = \begin{bmatrix} 0 & 0 & 0 & 0 & 0 & 0 \\ L_{2211} & L_{2222} & L_{2233} & 0 & 0 & 0 \\ L_{3311} & L_{3322} & L_{3333} & 0 & 0 & 0 \\ 0 & 0 & 0 & 2L_{2323} & 0 & 0 \\ 0 & 0 & 0 & 0 & 2L_{1313} & 0 \\ 0 & 0 & 0 & 0 & 0 & 2L_{1212} \end{bmatrix}, \quad \text{(A.2)}$$

$$L_{2211} = L_{3311} = \frac{\nu^m}{2(1-\nu^m)},\ L_{2222} = L_{3333} = \frac{1}{2(1-\nu^m)}\left[\frac{3}{4} + \frac{(1-2\nu^m)}{2}\right],$$
$$L_{1212} = L_{1313} = 1/4,$$
$$L_{2233} = L_{3322} = \frac{1}{2(1-\nu^m)}\left[\frac{1}{4} - \frac{(1-2\nu^m)}{2}\right],\ L_{2323} = \frac{1}{2(1-\nu^m)}\left[\frac{1}{4} + \frac{(1-2\nu^m)}{2}\right]. \quad \text{(A.3)}$$

ν^m is Poisson's ratio of the matrix.

A.2 Mori-Tanaka method

Non-zero bridging tensor elements of Mori-Tanaka method are given below [54]:

$$A_{11} = \frac{E^m}{E^f_{11}} \left(1 + \frac{\nu^m \left(\nu^m - \nu^f_{12}\right)}{(1 + \nu^m)(1 - \nu^m)}\right), \tag{A.4}$$

$$A_{12} = \frac{E^m}{E^f_{22}} \frac{\nu^m \left(1 - \nu^f_{23}\right)}{2(1 + \nu^m)(1 - \nu^m)} - \frac{E^m}{E^f_{11}} \frac{\nu^f_{12}}{(1 + \nu^m)(1 - \nu^m)} + \frac{\nu^m}{2(1 - \nu^m)} = A_{13}, \tag{A.5}$$

$$A_{21} = \frac{E^m}{E^f_{11}} \frac{\nu^m - \nu^f_{12}}{2(1 + \nu^m)(1 - \nu^m)} = A_{31}, \tag{A.6}$$

$$A_{22} = \frac{E^m}{E^f_{22}} \frac{\left(v^f_{23} - 3\right)}{8(v^m - 1)(v^m + 1)} + \frac{E^m}{E^f_{11}} \frac{v^f_{12} v^m}{2(v^m - 1)(v^m + 1)} + \frac{(v^m + 1)(4v^m - 5)}{8(v^m - 1)(v^m + 1)} = A_{33}, \tag{A.7}$$

$$A_{32} = \frac{E^m}{E^f_{22}} \frac{\left(3v^f_{23} - 1\right)}{8(v^m - 1)(v^m + 1)} + \frac{E^m}{E^f_{11}} \frac{v^f_{12} v^m}{2(v^m - 1)(v^m + 1)} + \frac{(v^m + 1)(1 - 4v^m)}{8(v^m - 1)(v^m + 1)} = A_{23}, \tag{A.8}$$

$$A_{44} = \frac{G^m}{G^f_{23}} \frac{1}{4(1 - v^m)} + \frac{(3 - 4v^m)}{4(1 - v^m)}, \tag{A.9}$$

$$A_{55} = \frac{G^m + G^f_{12}}{2G^f_{12}} = A_{66}. \tag{A.10}$$

G^f_{23} is the fiber transverse shear modulus.

A.3 Rule of mixture model

By this model, five elastic moduli of the composite are obtained as [19]

$$E_{11} = V_f E^f_{11} + V_m E^m, \tag{A.11}$$

$$\nu_{12} = V_f \nu^f_{12} + V_m \nu^m, \tag{A.12}$$

$$E_{22} = \frac{E^m}{1 - V_f\left(1 - E^m / E^f_{22}\right)}, \tag{A.13}$$

$$G_{12} = G_{13} = \frac{G^m}{1 - V_f\left(1 - G^m / G^f_{12}\right)}, \tag{A.14}$$

$$G_{23} = \frac{G^m}{1 - V_f\left(1 - G^m / G^f_{23}\right)}. \tag{A.15}$$

A.4 Chamis model

By simply replacing V_f in Eqs. (A.13)–(A.15) with $\sqrt{V_f}$, respectively, one obtains the Chamis model's formulae [20].

A.5 Modified rule of mixture model

Formulae for E_{11} and ν_{12} are the same as Eqs. (A.11) and (A.12). The modified formulae for the other three moduli are [19]

$$E_{22} = \frac{4\eta_{22}G_{23}}{\eta_{22} + mG_{23}},\ m = 1 + \frac{4\eta_{22}\nu_{12}^2}{E_{11}},\ \frac{1}{\eta_{22}} = \frac{1}{V_f + \eta_k V_m}\left(\frac{V_f}{\Lambda_{22}^f} + \frac{V_m \eta_k}{\Lambda^m}\right), \quad (A.16)$$

$$\frac{1}{G_{12}} = \frac{1}{V_f + \eta_{12} V_m}\left(\frac{V_f}{G_{12}^f} + \frac{V_m \eta_{12}}{G^m}\right),\ \eta_{12} = \frac{1}{2}\left(1 + \frac{G^m}{G_{12}^f}\right), \quad (A.17)$$

$$\frac{1}{G_{23}} = \frac{1}{V_f + \eta_{23} V_m}\left(\frac{V_f}{G_{23}^f} + \frac{V_m \eta_{23}}{G^m}\right),\ \eta_{23} = \frac{1}{4(1-\nu^m)}\left(3 - 4\nu^m + \frac{G^m}{G_{23}^f}\right), \quad (A.18)$$

$$\eta_k = \frac{1}{2(1-\nu^m)}\left(1 + \frac{\Lambda^m}{\Lambda_{22}^f}\right), \quad (A.19)$$

$$\Lambda_{22}^f = 0.5\left(K_{22}^f + K_{23}^f\right),\ \Lambda^m = 0.5\left(K_{22}^m + K_{23}^m\right). \quad (A.20)$$

K_{ij}^f and K_{ij}^m are the stiffness elements of the fiber and matrix.

A.6 Halpin-Tsai formulae

$$E_{22} = \frac{4\Lambda_L G_{23}}{\Lambda_L + mG_{23}},\ m = 1 + \frac{4\Lambda_L \nu_{12}^2}{E_{11}},\ \frac{\Lambda_L}{\Lambda^m} = \frac{1 + (1 - 2\nu^m)\eta V_f}{1 - \eta V_f}, \quad (A.21)$$

$$\frac{1}{G_{12}} = \frac{1}{V_f + \eta_{12} V_m}\left(\frac{V_f}{G_{12}^f} + \frac{V_m \eta_{12}}{G^m}\right),\ \eta_{12} = \frac{1}{2}\left(1 + \frac{G^m}{G_{12}^f}\right), \quad (A.22)$$

$$\frac{1}{G_{23}} = \frac{1}{V_f + \eta_{23} V_m}\left(\frac{V_f}{G_{23}^f} + \frac{V_m \eta_{23}}{G^m}\right),\ \eta_{23} = \frac{1}{4(1-\nu^m)}\left(3 - 4\nu^m + \frac{G^m}{G_{23}^f}\right), \quad (A.23)$$

$$\eta = \frac{\Lambda_{22}^f/\Lambda^m - 1}{\Lambda_{22}^f/\Lambda^m + 1 - 2\nu^m}. \quad (A.24)$$

It is noted that the formulae for E_{11} and ν_{12} are the same as Eqs. (A.11) and (A.12), respectively [21].

A.7 Hill-Hashin-Christensen-Lo model

$$E_{11} = V_f E_{11}^f + V_m E^m + \frac{4\left(\nu_{12}^f - \nu^m\right)^2 V_f(1 - V_f)}{\frac{V_f}{\Lambda^m} + \frac{1 - V_f}{\Lambda_{22}^f} + \frac{1}{G^m}}, \quad (A.25)$$

$$\nu_{12} = V_f \nu_{12}^f + V_m \nu^m + \frac{\left(\nu_{12}^f - \nu^m\right) V_f(1 - V_f)}{\frac{V_f}{\Lambda^m} + \frac{1 - V_f}{\Lambda_{22}^f} + \frac{1}{G^m}}\left(\frac{1}{\Lambda^m} - \frac{1}{\Lambda_{22}^f}\right), \quad (A.26)$$

$$E_{22} = \frac{2}{0.5/K_T + 0.5/G_{23} + 2\nu_{12}^2/E_{11}}, \tag{A.27}$$

$$G_{12} = G^m \frac{\left(G_{12}^f + G^m\right) + V_f\left(G_{12}^f - G^m\right)}{\left(G_{12}^f + G^m\right) - V_f\left(G_{12}^f - G^m\right)}, \tag{A.28}$$

$$K_T = K^m + \frac{V_f}{\frac{1}{\Lambda_{22}^f - \Lambda^m} + \frac{1-V_f}{\Lambda^m + G^m}}, \tag{A.29}$$

$$A_1\left(\frac{G_{23}}{G_m}\right)^2 + 2B\left(\frac{G_{23}}{G_m}\right) + C_1 = 0, \tag{A.30}$$

$$A_1 = 3V_f{V_m}^2\left(\frac{G^f}{G^m} - 1\right)\left(\frac{G^f}{G^m} + \eta_f\right) + \left[\frac{G^f}{G^m}\eta_m + \eta_f\eta_m - \left(\frac{G^f}{G^m}\eta_m - \eta_f\right){V_f}^3\right] \left[V_f\eta_m\left(\frac{G^f}{G^m} - 1\right) - \left(\frac{G^f}{G^m}\eta_m + 1\right)\right], \tag{A.31}$$

$$B_1 = -3V_f{V_m}^2\left(\frac{G^f}{G^m} - 1\right)\left(\frac{G^f}{G^m} + \eta_f\right) + \frac{1}{2}\left[\frac{G^f}{G^m}\eta_m + \left(\frac{G^f}{G^m} - 1\right)V_f + 1\right] \left[(\eta_m - 1)\left(\frac{G^f}{G^m} + \eta_f\right) - 2\left(\frac{G^f}{G^m}\eta_m - \eta_f\right){V_f}^3\right] + \frac{V_f}{2}(\eta_m + 1) \left(\frac{G^f}{G^m} - 1\right)\left[\frac{G^f}{G^m} + \eta_f + \left(\frac{G^f}{G^m}\eta_m - \eta_f\right){V_f}^3\right], \tag{A.32}$$

$$C_1 = 3V_f{V_m}^2\left(\frac{G^f}{G^m} - 1\right)\left(\frac{G^f}{G^m} + \eta_f\right) + \left[\frac{G^f}{G^m}\eta_m + V_f\left(\frac{G^f}{G^m} - 1\right) + 1\right] \left[\frac{G^f}{G^m} + \eta_f + \left(\frac{G^f}{G^m}\eta_m - \eta_f\right){V_f}^3\right], \tag{A.33}$$

$$\eta_f = 3 - 4\nu_{23}^f, \eta_m = 3 - 4\nu_m. \tag{A.34}$$

Eq. (A.30) is applicable only to composites with an isotropic fiber reinforcement [22].

A.8 Self-consistent model

The self-consistent formulae are represented as follows [15]:

$$[K_{ij}] = \left[K_{ij}^m\right] + V_f\left(\left[K_{ij}^f\right] - \left[K_{ij}^m\right]\right)\left[M_{ij}^f\right] \tag{A.35}$$

$$\left[M_{ij}^f\right] = \left\{[I] + \left[\tilde{L}_{ij}\right][S_{ij}]\left(\left[K_{ij}^f\right] - [K_{ij}]\right)\right\}^{-1} \tag{A.36}$$

$$\left[\tilde{L}_{ij}\right] = \left([S_{ij}]\left[\tilde{A}_{ij}\right]\left[K_{ij}^f\right] - [I]\right)\left(\left[K_{ij}^f\right] - [K_{ij}]\right)^{-1}[K_{ij}] \tag{A.37}$$

where $\left[\tilde{A}_{ij}\right]$ is the bridging tensor correlating the homogenized stresses of the composite with those of the fiber in a concentric cylinder assemblage (CCA) model, i.e., a two-phase CCA model, through $\{\sigma_i\} = \left[\tilde{A}_{ij}\right]\left\{\sigma_j^f\right\}$, whose non-zero elements are given by [54]

$$\tilde{A}_{11} = \frac{E_{11}\left(E_{11}\text{-}E_{22}\nu_{12}^f\nu_{12}\right)}{E_{11}^f\left[E_{11}\text{-}E_{22}(\nu_{12})^2\right]}, \tag{A.38}$$

$$\tilde{A}_{12} = \tilde{A}_{13} = \frac{E_{11}\left\{E_{11}^f E_{22}\left(1-\nu_{23}^f\right)\nu_{12} + E_{22}^f\left[\text{-}2E_{11}\nu_{12}^f + E_{11}^f\nu_{12}(1+\nu_{23})\right]\right\}}{2E_{11}^f E_{22}^f\left[E_{11} - E_{22}(\nu_{12})^2\right]}, \tag{A.39}$$

$$\tilde{A}_{21} = \tilde{A}_{31} = \frac{E_{11}E_{22}\left(\nu_{12} - \nu_{12}^f\right)}{2E_{11}^f\left[E_{11} - E_{22}(\nu_{12})^2\right]}, \tag{A.40}$$

$$\tilde{A}_{22} = \tilde{A}_{33} = \frac{E_{11}\left[E_{22}^f(5+\nu_{23}) + E_{22}\left(3\text{-}\nu_{23}^f\right)\right]}{8E_{22}^f\left[E_{11} - E_{22}(\nu_{12})^2\right]} - \frac{E_{22}\nu_{12}\left(E_{11}\nu_{12}^f + E_{11}^f\nu_{12}\right)}{2E_{11}^f\left[E_{11} - E_{22}(\nu_{12})^2\right]}, \tag{A.41}$$

$$\tilde{A}_{23} = \tilde{A}_{32} = \frac{E_{11}\left[E_{22}\left(1-3\nu_{23}^f\right) + E_{22}^f(3\nu_{23}\text{-}1)\right]}{8E_{22}^f\left[E_{11} - E_{22}(\nu_{12})^2\right]} + \frac{E_{22}\nu_{12}\left(E_{11}^f\nu_{12}\text{-}E_{11}\nu_{12}^f\right)}{2E_{11}^f\left[E_{11} - E_{22}(\nu_{12})^2\right]}, \tag{A.42}$$

$$\tilde{A}_{44} = \frac{\left(1+\nu_{23}^f\right)\left\{\text{-}8E_{22}G_{23}^f(\nu_{12})^2 + E_{11}\left[E_{22} + 2G_{23}^f(3-\nu_{23})\right]\right\}}{4E_{22}^f\left[E_{11} - E_{22}(\nu_{12})^2\right]}, \tag{A.43}$$

$$\tilde{A}_{55} = \tilde{A}_{66} = \frac{G_{12}^f + G_{12}}{2G_{12}^f}. \tag{A.44}$$

The self-consistent model is implicit, and an iteration has to be carried out to determine the five effective elastic moduli, E_{11}, E_{22}, G_{12}, ν_{12}, and ν_{23}, of the composite.

A.9 Generalized self-consistent model

Basic equations of the generalized self-consistent model are the same as Eqs. (A.35)–(A.37), except that the bridging tensor in Eq. (A.37), $\left[\tilde{A}_{ij}\right]$, correlating the stresses of the composite with those of the fiber is no longer obtained on a two-phase CCA model but on a three-phase CCA one shown in **Figure A.1**. In the latter case, however, generally non-explicit expressions exist for the bridging tensor elements $\tilde{A}_{ij}$. Solution for the resulting linear algebraic equations is necessary [54].

As pointed out in Ref. [54], a three-phase CCA model (**Figure A.1**) can be sufficiently well approximated with two step two-phase CCA ones. In the first step, the fiber and matrix phase constitutes a UD composite, whose effective elastic moduli, E_{11}^{eq}, E_{22}^{eq}, ν_{12}^{eq}, G_{12}^{eq}, and ν_{23}^{eq}, can be obtained from Eq. (6) in which the bridging tensor $[A_{ij}]$ is defined by Eqs. (A.4)–(A.10). Then, this UD composite is regarded as an equivalent fiber to embed into the composite. The resulting bridging tensor $\left[A_{ij}^{eq}\right]$ correlating the stresses of the composite with those of the equivalent fiber is given by Eqs. (A.38)–(A.44), providing that E_{11}^f, E_{22}^f, ν_{12}^f, G_{12}^f, and ν_{23}^f involved are replaced by E_{11}^{eq}, E_{22}^{eq}, ν_{12}^{eq}, G_{12}^{eq}, and ν_{23}^{eq}, respectively.

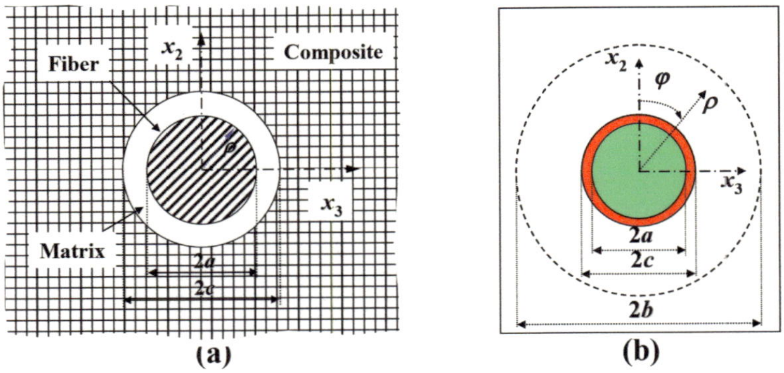

Figure A.1.
(a) Schematic for generalized self-consistent method. (b) A three-phase CCA model (b→) for the generalized self-consistent method.

From $\{\sigma_i\} = \left[A_{ij}^{eq}\right]\left\{\sigma_j^{eq}\right\}$ and $\left\{\sigma_i^f\right\} = \left(V_f[I] + V_m\left[A_{ij}\right]\right)^{-1}\left\{\sigma_j^{eq}\right\}$, it is obtained that

$$\left[\tilde{A}_{ij}\right] \approx \left[A_{ij}^{eq}\right]\left(V_f[I] + V_m\left[A_{ij}\right]\right). \tag{A.45}$$

A.10 Double inclusion method (DIM)

This method [4] has been incorporated into a commercial software, Digimat [40], for composites. As any prediction by this method is performed with Digimat, the formulae of the method are omitted.

Appendix B. Mechanical property data tables

	E-glass LY556	E-glass MY750	AS4 3501–6	T300 BSL914C	IM7 8511–7	T300 PR319	AS epoxy	S2-glass epoxy	G400–800 5260
E_{11}^f (GPa)	80	74	225	230	276	230	231	87	290
E_{22}^f (GPa)	80	74	15	15	19	15	15	87	19
ν_{12}^f	0.2	0.2	0.2	0.2	0.2	0.2	0.2	0.2	0.2
G_{12}^f (GPa)	33.33	30.8	15	15	27	15	15	36.3	27
ν_{23}^f	0.2	0.2	0.07	0.07	0.36	0.07	0.07	0.2	0.357
$\sigma_{u,t}^f$ (MPa)	2150	2150	3350	2500	5180	2500	3500	2850	5860
$\sigma_{u,c}^f$ (MPa)	1450	1450	2500	2000	3200	2000	3000	2450	3200
E^m (GPa)	3.35	3.35	4.2	4	4.08	0.95	3.2	3.2	3.45
ν^m	0.35	0.35	0.34	0.35	0.38	0.35	0.35	0.35	0.35
$\sigma_{u,t}^m$ (MPa)	80	80	69	75	99	70	85	73	70
$\sigma_{u,c}^m$ (MPa)	120	120	250	150	130	130	120	120	130
$\sigma_{u,s}^m$ (MPa)	54	54	50	70	57	41	50	52	57
V_f	0.62	0.6	0.6	0.6	0.6	0.6	0.6	0.6	0.6

Table B1.
Mechanical properties of the fibers and matrices of the nine UD composites used in WWFEs [25–27].

	E-glass LY556	E-glass MY750	AS4 3501–6	T300 BSL914C	IM7 8511–7	T300 PR319	AS epoxy	S2-glass epoxy	G400–800 5260
E_{11} (GPa)	53.5	45.6	126	138	165	129	140	52	173
E_{22} (GPa)	17.7	16.2	11	11	8.4	5.6	10	19	10
ν_{12}	0.278	0.278	0.28	0.28	0.34	0.318	0.3	0.3	0.33
G_{12} (GPa)	5.83	5.83	6.6	5.5	5.6	1.33	6	6.7	6.94
G_{23} (GPa)	6.32	5.79	3.93	3.93	2.8	1.86	3.35	6.7	3.56
$\sigma_{11}^{u,t}$ (MPa)	1140	1280	1950	1500	2560	1378	1990	1700	2750
$\sigma_{11}^{u,c}$ (MPa)	570	800	1480	900	1590	950	1500	1150	1700
$\sigma_{22}^{u,t}$ (MPa)	35	40	48	27	73	40	38	63	75
$\sigma_{22}^{u,c}$ (MPa)	114	145	200	200	185	125	150	180	210
σ_{12}^{u} (MPa)	72	73	79	80	90	97	70	72	90
σ_{23}^{u} (MPa)	50	50	55	-	57	45	50	40	57

Table B2.
Mechanical properties of the nine UD composites used in WWFEs [25–27].

	E-glass LY556	E-glass MY750	AS4 3501–6	T300 BSL914C	IM7 8511–7	T300 PR319	AS epoxy	S2-glass epoxy	G400–800 5260
				By Eshelby's method					
E_{11} (GPa)	50.81	45.68	136.61	139.51	167.07	138.36	139.81	53.42	175.31
E_{22} (GPa)	7.15	7.01	7.30	7.08	7.73	1.98	5.91	6.76	6.52
ν_{12}	0.28	0.28	0.27	0.27	0.29	0.28	0.27	0.28	0.28
G_{12} (GPa)	2.67	2.61	3.09	2.94	3.07	0.76	2.40	2.52	2.68
G_{23} (GPa)	2.42	2.37	2.60	2.49	2.50	0.67	2.05	2.28	2.19
				By Bridging Model					
E_{11} (GPa)	50.9	45.7	136.7	139.6	167.2	138.4	139.9	53.5	175.4
E_{22} (GPa)	18.1	16.8	9.7	9.6	11.2	4.41	8.7	16.9	10.2
ν_{12}	0.257	0.26	0.256	0.26	0.272	0.26	0.26	0.26	0.26
G_{12} (GPa)	6.28	5.84	5.54	5.35	6.46	1.82	4.64	5.81	5.8
G_{23} (GPa)	6.24	5.8	3.76	3.66	3.76	1.55	3.29	5.77	3.51
				By Mori-Tanaka's method					
E_{11} (GPa)	50.9	45.76	136.7	139.6	167.3	138.4	139.9	53.5	175.4
E_{22} (GPa)	11.7	11.02	8.757	8.573	9.665	3.02	7.481	10.78	8.473
ν_{12}	0.249	0.252	0.26	0.257	0.267	0.252	0.256	0.252	0.254
G_{12} (GPa)	4.60	4.32	4.53	4.35	4.92	1.30	3.67	4.23	4.36
G_{23} (GPa)	4.06	3.83	3.32	3.21	3.23	1.06	2.77	3.72	2.92
				By rule of mixture method					
E_{11} (GPa)	50.87	45.74	136.7	139.6	167.2	138.38	139.88	53.5	175.4
E_{22} (GPa)	8.252	7.84	7.394	7.14	7.715	2.169	6.061	7.5817	6.779
ν_{12}	0.257	0.26	0.256	0.26	0.272	0.26	0.26	0.26	0.26

	E-glass LY556	E-glass MY750	AS4 3501–6	T300 BSL914C	IM7 8511–7	T300 PR319	AS epoxy	S2-glass epoxy	G400–800 5260
G_{12} (GPa)	3.08	2.92	3.39	3.23	3.42	0.85	2.65	2.82	2.99
G_{23} (GPa)	3.08	2.92	2.93	2.81	2.81	0.82	2.36	2.82	2.51
By Chamis method									
E_{11} (GPa)	50.87	45.74	136.7	139.6	167.2	138.38	139.88	53.5	175.4
E_{22} (GPa)	13.64	12.86	9.496	9.26	10.415	3.461	8.192	12.604	9.425
ν_{12}	0.257	0.26	0.256	0.26	0.272	0.272	0.26	0.26	0.26
G_{12} (GPa)	5.13	4.83	5.12	4.91	5.52	1.45	4.14	4.73	4.88
G_{23} (GPa)	5.13	4.83	3.93	3.81	3.80	1.33	3.33	4.73	3.49
By modified rule of mixture method									
E_{11} (GPa)	50.87	45.74	136.7	139.6	167.2	138.38	139.88	53.5	175.4
E_{22} (GPa)	11.61	10.93	8.65	8.46	9.53	2.98	7.37	10.71	8.35
ν_{12}	0.257	0.26	0.256	0.26	0.272	0.272	0.26	0.26	0.26
G_{12} (GPa)	4.60	4.32	4.54	4.35	4.92	1.29	3.67	4.23	4.35
G_{23} (GPa)	4.06	3.82	3.32	3.21	3.23	1.06	2.77	3.73	2.91
By Halpin-Tsai formulae									
E_{11} (GPa)	50.87	45.74	136.7	139.6	167.2	138.38	139.88	53.5	175.4
E_{22} (GPa)	11.69	11.0	8.76	8.57	9.66	3.02	7.48	10.77	8.47
ν_{12}	0.257	0.26	0.256	0.26	0.272	0.272	0.26	0.26	0.26
G_{12} (GPa)	4.60	4.32	4.54	4.35	4.92	1.29	3.67	4.23	4.35
G_{23} (GPa)	4.06	3.82	3.32	3.21	3.23	1.06	2.77	3.73	2.91
By Hill-Hashin-Christensen-Lo method									
E_{11} (GPa)	50.9	45.8	136.7	139.6	167.3	138.4	139.9	53.5	175.4

	E-glass LY556	E-glass MY750	AS4 3501–6	T300 BSL914C	IM7 8511–7	T300 PR319	AS epoxy	S2-glass epoxy	G400–800 5260
E_{22} (GPa)	12.9	12.0	—	—	—	—	—	11.85	—
ν_{12}	0.249	0.252	0.253	0.257	0.267	0.252	0.256	0.25	0.25
G_{12} (GPa)	4.6	4.32	4.54	4.35	4.92	1.29	3.67	4.23	4.35
G_{23} (GPa)	4.65	4.33	—	—	—	—	—	4.25	—
By self-consistent method									
E_{11} (GPa)	50.94	45.81	136.72	139.65	167.31	138.40	139.92	53.55	175.43
E_{22} (GPa)	18.91	16.80	9.14	8.99	10.37	4.19	8.06	17.39	9.33
ν_{12}	0.231	0.235	0.250	0.254	0.262	0.238	0.251	0.233	0.247
G_{12} (GPa)	11.34	9.80	6.37	6.25	9.36	4.18	5.82	10.94	8.98
G_{23} (GPa)	6.96	6.15	3.53	3.43	3.52	1.55	3.06	6.36	3.25
By generalized self-consistent method									
E_{11} (GPa)	50.9	45.8	136.7	139.6	167.3	138.4	139.9	53.5	175.4
E_{22} (GPa)	12.87	12.03	8.93	8.77	10.1	3.27	7.72	11.8	8.85
ν_{12}	0.249	0.252	0.253	0.257	0.27	0.25	0.256	0.25	0.254
G_{12} (GPa)	4.6	4.32	4.54	4.3	4.9	1.2	3.6	4.2	4.35
G_{23} (GPa)	4.65	4.33	3.42	3.32	3.42	1.19	2.9	4.25	3.09
By double-inclusion method (Digimat [40])									
E_{11} (GPa)	50.9	47.2	141.1	144.2	172.8	143	144.5	55.2	181.2
E_{22} (GPa)	16.2	15.9	9.35	9.2	10.5	4.11	8.26	16.1	9.48
ν_{12}	0.234	0.234	0.248	0.252	0.257	0.238	0.249	0.234	0.244
G_{12} (GPa)	6.73	6.585	5.8	5.63	7.12	2.13	5	6.65	6.46

	E-glass LY556	E-glass MY750	AS4 3501–6	T300 BSL914C	IM7 8511–7	T300 PR319	AS epoxy	S2-glass epoxy	G400–800 5260
G_{23} (GPa)	5.78	5.67	3.64	3.54	3.55	1.51	3.16	5.7	3.28
By FE-square fiber array									
E_{11} (GPa)	50.9	45.8	136.7	139.6	167.3	138.4	139.9	53.5	175.4
E_{22} (GPa)	16.26	14.9	9.54	9.42	10.88	3.98	8.45	14.86	9.63
ν_{12}	0.246	0.25	0.252	0.256	0.266	0.25	0.255	0.249	0.252
G_{12} (GPa)	4.96	4.58	4.68	4.5	5.15	1.38	3.82	4.5	4.57
G_{23} (GPa)	6.49	5.89	3.79	3.71	3.79	1.58	3.32	5.89	3.47
By FE-hexagonal fiber array									
E_{11} (GPa)	50.89	45.77	136.70	139.56	167.26	138.38	139.90	53.50	175.37
E_{22} (GPa)	12.60	11.72	8.89	8.72	9.89	3.19	7.66	11.51	8.68
ν_{12}	0.249	0.251	0.253	0.255	0.267	0.259	0.256	0.251	0.252
G_{12} (GPa)	4.62	4.32	4.54	4.35	4.91	1.30	3.67	4.22	4.35
G_{23} (GPa)	3.03	4.18	3.40	3.29	3.32	1.14	2.86	4.07	3.01
By FE-square diagonal fiber array									
E_{11} (GPa)	50.89	45.77	136.71	139.56	167.30	138.42	139.88	53.51	175.43
E_{22} (GPa)	10.40	9.85	8.30	8.09	9.03	2.71	7.12	9.60	7.92
ν_{12}	0.248	0.250	0.251	0.255	0.264	0.251	0.260	0.250	0.255
G_{12}(GPa)	4.87	4.56	4.68	4.50	5.16	1.38	3.82	4.48	4.58
G_{23} (GPa)	2.59	3.25	3.06	2.94	2.95	0.90	2.50	3.15	2.65
By FE-random fiber array									
E_{11} (GPa)	50.90	45.77	136.71	139.37	167.27	138.37	139.90	53.51	175.41
E_{22} (GPa)	13.76	12.79	8.92	8.76	9.97	3.43	7.73	12.77	8.84

	E-glass LY556	E-glass MY750	AS4 3501–6	T300 BSL914C	IM7 8511–7	T300 PR319	AS epoxy	S2-glass epoxy	G400–800 5260
ν_{12}	0.241	0.245	0.248	0.255	0.263	0.241	0.252	0.243	0.250
G_{12} (GPa)	4.58	4.29	4.42	4.33	4.69	1.29	3.60	4.21	4.29
G_{23} (GPa)	5.08	4.74	3.48	3.53	3.43	1.28	2.97	4.69	3.12

Table B3.
Predicted elastic moduli of the nine UD composites by different models.

DOI: http://dx.doi.org/10.5772/intechopen.80807

	E-glass LY556	E-glass MY750	AS4 3501–6	T300 BSL914C	IM7 8511–7	T300 PR319	AS epoxy	S2-glass epoxy	G400–800 5260
Eshelby's method									
λ_1	1.577	1.623	1.648	1.650	1.654	1.663	1.653	1.631	1.655
λ_2	1.370	1.322	1.204	1.205	1.247	1.308	1.234	1.324	1.270
λ_3	1.301	1.257	1.220	1.218	1.201	1.255	1.228	1.259	1.225
λ_4	1.169	1.131	1.123	1.124	1.130	1.133	1.126	1.132	1.130
Bridging Model									
λ_1	1.573	1.618	1.646	1.648	1.650	1.662	1.651	1.627	1.654
λ_2	0.442	0.453	0.621	0.612	0.577	0.467	0.576	0.446	0.554
λ_3	0.442	0.453	0.621	0.612	0.577	0.467	0.576	0.446	0.554
λ_4	0.438	0.449	0.498	0.494	0.460	0.436	0.479	0.443	0.454
Mori-Tanaka's method									
λ_1	1.571	1.616	1.647	1.647	1.650	1.662	1.651	1.625	1.653
λ_2	0.793	0.801	0.868	0.867	0.855	0.807	0.852	0.797	0.841
λ_3	0.734	0.739	0.800	0.793	0.782	0.743	0.780	0.738	0.782
λ_4	0.635	0.644	0.674	0.670	0.650	0.636	0.661	0.640	0.647
Rule of mixture method									
λ_1	1.573	1.618	1.646	1.648	1.650	1.662	1.651	1.627	1.654
λ_2	1.304	1.303	1.661	1.755	1.725	1.352	1.576	1.304	1.418
λ_3	1.001	1.002	1.000	1.001	1.000	0.999	1.000	1.000	0.998
λ_4	1.000	1.001	1.000	1.000	1.000	1.000	1.000	1.000	1.000
Chamis method									
λ_1	1.573	1.618	1.646	1.648	1.650	1.662	1.651	1.627	1.654

	E-glass LY556	E-glass MY750	AS4 3501–6	T300 BSL914C	IM7 8511–7	T300 PR319	AS epoxy	S2-glass epoxy	G400–800 5260
λ_2	0.720	0.724	0.925	0.977	0.964	0.761	0.882	0.726	0.796
λ_3	0.560	0.564	0.564	0.564	0.564	0.563	0.564	0.564	0.562
λ_4	0.560	0.564	0.563	0.563	0.564	0.564	0.563	0.564	0.564
Modified rule of mixture method									
λ_1	1.573	1.618	1.646	1.648	1.650	1.662	1.651	1.627	1.654
λ_2	0.808	0.813	0.957	0.966	0.934	0.840	0.933	0.811	0.890
λ_3	0.734	0.741	0.800	0.793	0.780	0.742	0.779	0.737	0.785
λ_4	0.635	0.643	0.672	0.671	0.650	0.638	0.662	0.640	0.648
Halpin-Tsai formulae									
λ_1	1.573	1.618	1.646	1.648	1.650	1.662	1.651	1.627	1.654
λ_2	0.794	0.800	0.867	0.870	0.860	0.809	0.856	0.800	0.839
λ_3	0.734	0.741	0.800	0.793	0.780	0.742	0.779	0.737	0.785
λ_4	0.635	0.643	0.672	0.671	0.650	0.638	0.662	0.640	0.648
Hill-Hashin-Christensen-Lo method									
λ_1	1.571	1.616	—	—	—	—	—	1.625	—
λ_2	0.727	0.755	—	—	—	—	—	0.745	—
λ_3	0.636	0.642	—	—	—	—	—	0.636	—
λ_4	0.635	0.643	—	—	—	—	—	0.640	—
Self-consistent method									
λ_1	1.567	1.612	1.645	1.646	1.649	1.662	1.650	1.622	1.653
λ_2	0.459	0.496	0.806	0.795	0.764	0.549	0.755	0.471	0.713

	E-glass LY556	E-glass MY750	AS4 3501-6	T300 BSL914C	IM7 8511-7	T300 PR319	AS epoxy	S2-glass epoxy	G400-800 5260
λ_3	0.385	0.420	0.710	0.699	0.661	0.465	0.657	0.396	0.644
λ_4	0.197	0.224	0.395	0.382	0.272	0.155	0.338	0.195	0.250
Generalized self-consistent method									
λ_1	1.571	1.616	1.646	1.647	1.650	1.662	1.651	1.625	1.653
λ_2	0.792	0.799	0.870	0.867	0.854	0.809	0.857	0.798	0.839
λ_3	0.734	0.740	0.800	0.793	0.780	0.742	0.779	0.738	0.785
λ_4	0.635	0.643	0.673	0.671	0.650	0.638	0.662	0.640	0.648
Double inclusion method (Digimat)									
λ_1	1.568	1.617	1.647	1.648	1.650	1.662	1.652	1.626	1.654
λ_2	0.543	0.517	0.766	0.758	0.729	0.559	0.721	0.502	0.679
λ_3	0.485	0.465	0.666	0.657	0.649	0.481	0.620	0.453	0.633
λ_4	0.402	0.386	0.463	0.456	0.404	0.363	0.429	0.376	0.396
FE-square fiber array									
λ_1	1.570	1.616	1.645	1.647	1.649	1.662	1.651	1.625	1.653
λ_2	0.616	0.635	0.773	0.769	0.744	0.650	0.735	0.630	0.721
λ_3	0.421	0.444	0.612	0.596	0.566	0.454	0.565	0.436	0.568
λ_4	0.582	0.600	0.643	0.639	0.614	0.593	0.628	0.596	0.611
FE-hexagonal fiber array									
λ_1	1.571	1.616	1.646	1.646	1.650	1.662	1.651	1.625	1.653
λ_2	0.295	0.766	0.853	0.850	0.825	0.775	0.828	0.760	0.815
λ_3	1.019	0.669	0.765	0.756	0.741	0.677	0.737	0.667	0.742
λ_4	0.633	0.643	0.673	0.671	0.651	0.636	0.662	0.641	0.647

	E-glass LY556	E-glass MY750	AS4 3501-6	T300 BSL914C	IM7 8511-7	T300 PR319	AS epoxy	S2-glass epoxy	G400-800 5260
				FE- FE-square diagonal fiber array					
λ_1	1.571	1.616	1.645	1.646	1.649	1.662	1.651	1.625	1.653
λ_2	0.454	0.856	0.932	0.928	0.921	0.864	0.809	0.856	0.898
λ_3	1.210	0.890	0.928	0.926	0.919	0.893	0.918	0.887	0.916
λ_4	0.594	0.604	0.643	0.640	0.613	0.593	0.628	0.599	0.609
				FE-random fiber array					
λ_1	1.570	1.615	1.645	1.647	1.649	1.662	1.651	1.625	1.653
λ_2	0.660	0.676	0.830	0.826	0.802	0.694	0.798	0.662	0.777
λ_3	0.566	0.578	0.731	0.662	0.695	0.592	0.692	0.569	0.693
λ_4	0.611	0.621	0.606	0.601	0.634	0.609	0.650	0.615	0.629

Table B4.
The coefficients λ_i's for the nine composites by different models.

Author details

Zheng-Ming Huang
School of Aerospace Engineering and Applied Mechanics, Tongji University, Shanghai, China

*Address all correspondence to: huangzm@tongji.edu.cn

References

[1] Llorca J, González C, Molina-Aldareguía JM, Segurado J, Seltzer R, Sket F, et al. Multiscale modeling of composite materials: A roadmap towards virtual testing. Advanced Materials. 2011;**23**:5130-5147

[2] Bogdanovich AE, Sierakowski RL. Composite materials and structures: Science, technology and applications—A compendium of books, review papers, and other sources of information. Applied Mechanics Reviews. 1999; **52**(12):351-366

[3] Pindera MJ, Khatam H, Drago AS, Bansal Y. Micromechanics of spatially uniform heterogeneous media: A critical review and emerging approaches. Composites, Part B. 2009;**40**(5):349-378

[4] Nemat-Nasser S, Hori M. Micromechanics: Overall Properties of Heterogeneous Materials. Amsterdam: Elsevier; 2013

[5] Hinton MJ, Kaddour AS, Soden PD. Failure Criteria in Fiber Reinforced Polymer Composites. Amsterdam, The Netherlands: Elsevier; 2004

[6] Eshelby JD. The determination of the elastic field of an ellipsoidal inclusion and related Problems. Proceedings of the Royal Society. 1957;**A240**:367-396

[7] Chen T, Dvorak GJ, Benveniste Y. Stress fields in composites reinforced by coated cylindrically orthotropic fibers. Mechanics of Materials. 1990;**9**:17-32

[8] Huang Z-M, Liu L. Predicting strength of fibrous laminates under triaxial loads only upon independently measured constituent properties. International Journal of Mechanical Sciences. 2014;**79**(1):105-129

[9] Huang Z-M, Xin L-M. Stress concentration factor in matrix of a composite subjected to transverse compression. International Journal of Applied Mechanics. 2016;**8**(3): 1650034-1-1650034-165003418

[10] Huang Z-M, Xin L-M. In situ strengths of matrix in a composite. Acta Mechanica Sinica. 2017;**33**(1): 120-131

[11] Zhou Y, Huang Z-M, Liu L. Prediction of interfacial debonding in fiber-reinforced composite laminates. Polymer Composites. (Accepted for publication)

[12] Ryan S, Wicklein M, Mouritz A, Riedel W, Schäfer F, Thoma K. Theoretical prediction of dynamic composite material properties for hypervelocity impact simulations. International Journal of Impact Engineering. 2009;**36**(7):899-912

[13] Younes R, Hallal A, Fardoun F, Chehade FH. Comparative review study on elastic properties modeling for unidirectional composite materials. In: Hu N, editor. Composites and Their Properties. InTech; 2012. pp. 391-408. DOI: 10.5772/50362

[14] Medikonda STA, Hamm R. A comparative study of the effect of representative volume cell (RVC) boundary conditions on the elastic properties of a micromechanics based unidirectional composite material model. International Journal of Computational Materials. 2017;7(2):51-71

[15] Qu J, Cherkaoui M. Fundamentals of Micromechanics of Solids. New Jersey: John Wiley & Sons; 2006

[16] Huang Z-M, Zhou Y-X. Strength of Fibrous Composites. Zhejiang Univ. New York: Press & Springer; 2011

[17] Mori T, Tanaka K. Average stress in matrix and average energy of materials

with misfitting inclusion. Acta Metallurgica. 1973;**21**:571-574

[18] Benveniste Y. A new approach to the application of Mori-Tanaka's theory in composite materials. Mechanics of Materials. 1987;**6**(2):147-157

[19] Tsai SW, Hahn HT. Introduction to Composite Materials. Bosa Roca: CRC Press; 1980

[20] Chamis CC. Mechanics of composite materials: Past, present and future. Journal of Composites Technology and Research. 1989;**11**(1):3-14

[21] Halpin JC, Kardos JL. The Halpin-Tsai equations: A review. Polymer Engineering and Science. 1976;**16**:344-352

[22] Swanson SR. Introduction to Design and Analysis with Advanced Composite Materials. New Jersey: Prentice Hall; 1997

[23] Christensen RM, Lo KH. Solutions for effective shear properties in three phase sphere and cylinder models. Journal of the Mechanics and Physics of Solids. 1979;**27**:315-330

[24] Barbero EJ. Finite Element Analysis of Composite Materials Using Abaqus. Bosa Roca: CRC Press; 2013

[25] Soden PD, Hinton MJ, Kaddour AS. Lamina properties, lay-up configurations and loading conditions for a range of fiber-reinforced composite laminates. Composites Science and Technology. 1998;**58**: 1011-1022

[26] Kaddour AS, Hinton JM. Input data for test cases used in benchmarking triaxial failure theories of composites. Journal of Composite Materials. 2012; **46**:2295-2312

[27] Kaddour AS et al. Mechanical properties and details of composite laminates for test cases used in the third world-wide failure exercise. Journal of Composite Materials. 2013; **47**:2427-2442

[28] Segurado J, Llorca J. A numerical approximation to the elastic properties of sphere-reinforced composites. Journal of the Mechanics and Physics of Solids. 2002;**50**(10):2107-2121

[29] Melro AR, Camanho PP, Pinho ST. Generation of random distribution of fibres in long-fibre reinforced composites. Composites Science and Technology. 2008;**68**(9):2092-2102

[30] Chateau C, Gelebart L, Bornert M, Crépinc J. Micromechanical modeling of the elastic behavior of unidirectional CVI SiC/SiC composites. International Journal of Solids and Structures. 2015; **58**:322-334

[31] Gonzalez C, LLorca J. Mechanical behavior of unidirectional fiber-reinforced polymers under transverse compression: Microscopic mechanisms and modeling. Composites Science and Technology. 2007;**67**(13):2795-2806

[32] Jimenez FL, Pellegrino S. Constitutive modeling of fiber composites with a soft hyperelastic matrix. International Journal of Solids and Structures. 2012;**49**(3–4):635-647

[33] Huang Y, Jin KK, Ha SK. Effects of fiber arrangement on mechanical behavior of unidirectional composites. Journal of Composite Materials. 2008; **42**(18):1851-1871

[34] Cavalcante MAA, Pindera M-J, Khatam H. Finite-volume micromechanics of periodic materials: Past, present and future. Composites, Part B. 2012;**43**:2521-2543

[35] Aboudi J, Arnold SM, Bednarcyk BA. Micromechanics of Composite Materials: A Generalized Multiscale Analysis Approach. Oxford: Butterworth-Heinemann; 2012

[36] Sertse HM, Goodsell J, Ritchey AJ, Pipes RB, Yu W. Challenge problems for the benchmarking of micromechanics analysis: Level I initial results. Journal of Composite Materials. 2018;**52**(1):61-80

[37] Wang YC, Huang Z-M. A new approach to a bridging tensor. Polymer Composites. 2015;**36**(8):1417-1431

[38] Sun CT, Vaidya RS. Prediction of composite properties from a representative volume element. Composites Science and Technology. 1996;**56**:171-179

[39] Tsai JL, Chi YK. Investigating thermal residual stress effect on mechanical behaviors of fiber composites with different fiber arrays. Composites, Part B. 2008;**39**(4):714-721

[40] DIGIMAT User Manual, Release 5.1.2, MSC Software Company; 2014

[41] Benveniste Y, Dvorak GJ, Chen T. Stress fields in composites with coated inclusions. Mechanics of Materials. 1989; 7:305-317

[42] Hobbiebrunken T, Hojo M, Adachi T, Jong CD, Fiedler B. Evaluation of interfacial strength in CF/epoxies using FEM and in-situ experiments. Composites, Part A. 2006;**37**(12): 2248-2256

[43] Mortell DJ, Tanner DA, McCarthy CT. In-situ SEM study of transverse cracking and delamination in laminated composite materials. Composites Science and Technology. 2014;**105**:118-126

[44] Koyanagi J, Ogihara S, Nakatani H, Okabe T, Yoneyama S. Mechanical properties of fiber/matrix interface in polymer matrix composites. Advanced Composite Materials. 2014;**23**(5–6): 551-570

[45] Budiansky B, Fleck NA. Compressive failure of fiber composites. Journal of the Mechanics and Physics of Solids. 1993;**41**:183-211

[46] Kyriakides S, Arseculeratne R, Perry EJ, Liechti KM. On the compressive failure of fiber reinforced composites. International Journal of Solids and Structures. 1995;**32**:689-738

[47] Basu S, Waas AM, Ambur DR. Compressive failure of fiber composites under multi-axial loading. Journal of the Mechanics and Physics of Solids. 2006; **54**:611-634

[48] Zhou Y, Huang Z-M. Composite failure under longitudinal compression. Submitted for publication

[49] ParíS F, Correa E, Cañas J. Micromechanical view of failure of the matrix in fibrous composite materials. Composites Science and Technology. 2003;**63**(7):1041-1052

[50] Wisnom MR. Factors affecting the transverse tensile strength of unidirectional continuous silicon carbide fibre reinforced 6061 aluminum. Journal of Composite Materials. 1990;**24**(7):707-726

[51] Aboudi J. Micromechanical analysis of the strength of unidirectional fiber composites. Composites Science and Technology. 1988;**33**(2):79-96

[52] Mayes S, Andrew C, Hansen J. Multicontinuum failure analysis of composite structural laminates. Mechanics of Composite Materials and Structures. 2001;**8**(4):249-262

[53] Pindera MJ, Gurdal Z, Hidde JS, Herakovich CT. Mechanical and Thermal Characterization of Unidirectional Aramid/Epoxy, CCMS-86-08, VPI-86-29. Virginia Polytechnic Institute and State University; 1986

[54] Wang YC, Huang Z-M. Bridging tensor with an imperfect interface. European Journal of Mechanics - A/Solids. 2016;**56**:73-91

Chapter 5

Probabilistic Modeling of Failure

Alireda Aljaroudi

Abstract

Failure of a system or a component of a system is and has been a major concern to systems' operators and owners. Failure could be traced back to different causes and may take different forms and shapes. It may result from software malfunction, hardware degraded performance, human error, sabotage, environmental as well as other external factors. There are various techniques found in the literature that can assist in the analysis of failure. These techniques comprise deterministic and probabilistic techniques. Deterministic techniques ignore the variability and uncertainties of the variables in the analysis which may lead to unsatisfactory and inaccurate results. While probabilistic techniques produce accurate and an all-inclusive result because they incorporate the variabilities and uncertainties in the analysis. The focus of this chapter is to present commonly used probabilistic failure analysis techniques and their mathematical derivations. Examples to enhance the understanding of the concept of failure analysis are also presented.

Keywords: failure analysis, probabilistic methods in engineering, reliability engineering

1. Introduction

Traditionally, failure analysis is conducted using deterministic techniques to assess the operability and integrity of industrial systems. These techniques lack the ability to report or predict the probabilistic nature of the systems' behavior. Moreover, they ignore the probabilistic and random nature of the external factors that have direct impact on the performance of the systems. Implementing these techniques may produce inadequate assessment and eventually, may lead to wrong decisions concerning the integrity and reliability of the evaluated systems. To make an informed and reliable decision about the reliability and operability of such systems, probabilistic failure analysis should be adopted as an alternative analysis technique. This technique should be made as an integral part of the decision process as well as be a part of the overall organization's risk control.

This chapter presents techniques that assist in the analysis of failure using engineering probabilistic methods. They include simulation as well as analytical methods. Simulation methods can be conducted using Monte Carlo simulation technique. Two different Monte Carlo simulation approaches are presented in this chapter. These are, the counting approach and sample statistics approach. The main drawback with simulation is that it takes great deal of time to perform and may require an extensive processing power. However, it is an essential step in the analysis to validate the results obtained by the analytical methods.

Some of the analytical methods include first order reliability method (FORM) and second order reliability method (SORM). FORM involves two approaches to calculate probability of failure, these are first order and second moment (FOSM) and advanced FOSM.

The focus of this chapter will be only on the FORM with the assumption that all random variables are uncorrelated. Analysis that require the use of FORM for correlated random variables is beyond the scope of this chapter. Likewise, analysis requiring the application of SORM to analyze limit state functions involving second order representation is beyond the scope of this chapter.

2. Failure modeling

Failure can be defined as the inability of industrial systems or subsystems either partially or totally to satisfy operational requirements as set forth by design specifications. Failure of a system could be partial or complete; in either case the consequences of failure may result in adverse consequences. Interruption of services, degraded performance, system shutdown, environmental damage and customers' dissatisfaction are some of the consequences. Such consequences may lead to financial losses, liabilities and destroyed image of the operating company. As an example, if failure involves leak detection system to detect oil and gas leakage from subsea oil and gas pipelines, consequences could be so severe. Pollution of the ocean, damage to the fishery and tourism industries are some of the major consequences.

The system fails when the imposed demand or load on the system exceeds its capacity or resistance. The strength or the capacity of the system is a design parameter that specifies the maximum load the system can endure or the maximum demand the system can satisfy. The variabilities of the system's capacity to satisfy the demand or load imposed on it are mainly attributed to the inherent uncertainties of the operation characteristics of the system's components as well as external environmental factors. Therefore, the capacity of the system is assumed to be probabilistic in nature that varies from time to time due to the reasons mentioned above. Likewise, using same argument the load or demand imposed on the system are considered probabilistic in nature due to the effect of the varying environmental conditions.

Considering the above, the performance function of the system or sometime is called limit state function can be formulated as the difference between the system's capacity and the load or demand imposed on it. The same argument can be used for production facility, the performance is the difference between supply and demand, supply being the capacity, or the strength and demand is the load. If the two parameters are the same then it can be said that the system is at a limit state, if the system cannot meet the demand then the system is at a failure state, and if the system capacity exceeds the load imposed on it, the system is at a satisfactory state.

3. Reliability analysis methods

Knowing in advance when the system is going to fail or degrade in performance is an essential step in the failure analysis. Under this step, the probability of failure is calculated in terms of the random variables affecting the performance of the system. There are several approaches found in the literature that can be used to evaluate the probability of failure either analytically or by simulation. Analytical methods approximate the probability of failure by using first order reliability method (FORM) or second order reliability method (SORM). The FORM uses two approximation techniques that evaluates the probability of failure, these are the first order and second moment (FOSM) and advanced first order second moment (AFOSM) techniques.

Calculating the probability of failure based on the methods mentioned above can be used to predict the ability of the system to satisfy operational as well as safety requirements during its life cycle. Combining this analysis with risk analysis, the consequences of failure can be easily determined. First order reliability method consists of two techniques namely:

3.1 Analytical methods

3.1.1 First order second moment (FOSM)

FOSM makes use of second moment statistics (mean and variance) and ignores higher moments (skewness and kurtosis) of the random variables. It evaluates the performance function by using the first order Taylor series expansion of the limit state function (LSF) at the mean value. This method is used when the performance function is linear having statistically independent, normally distributed and noncorrelated random variables X_i's.

Performance function can be defined as [3, 5, 10]:

$$Z = C - D \tag{1}$$

where C is the capacity and D is the demand are statistically independent random variables and are assumed to be normally distributed. Failure occurs when:

$$Z < 0 \tag{2}$$

Then the probability of failure (P_f) can be computed as:

$$P_f = (P(C < D)) \tag{3}$$

Or

$$P_f = (P(Z < 0)) \tag{4}$$

Figure 1 shows the probability density function (*PDF*) of the random variable Z, as it can be depicted from the figure that the probability of failure comprises the shaded area where $Z < 0$.

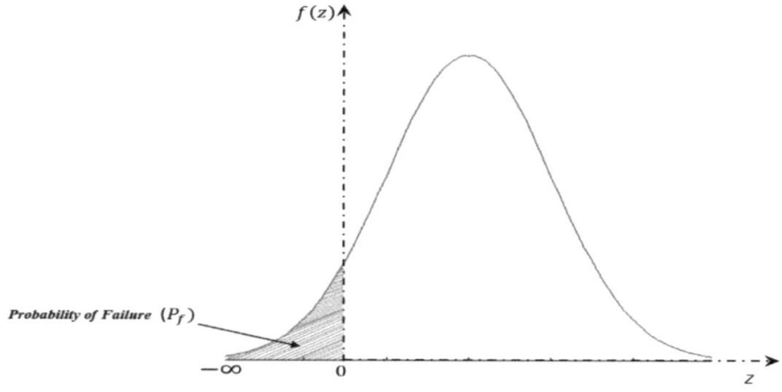

Figure 1.
Probability density function of the performance function.

The probability of failure is expressed as [3, 5, 10]:

$$P_f = (P(Z < 0)) = \int_{-\infty}^{0} f_{z(Z)dz} \tag{5}$$

Alternatively, the performance function Z can be formulated in terms of many random variables designated by a vector $\boldsymbol{X}$ as [3, 5, 10]:

$$Z(\boldsymbol{X}) = Z(X_1, X_2, \dots X_n) \tag{6}$$

$X_1, \dots X_n$ are the random variables in the performance function.

The integration of the performance function as indicated in Eq. (5) is performed for the region where $Z < 0$, this type of integration is difficult to solve, alternatively Taylor series expansion is used. The first order Taylor series approximation about the mean of the random variables is shown in Eq. (7). The expansion is truncated at the linear terms to obtain the first order approximation of the performance function [7–9].

$$Z = Z(\mu_X) + \sum_{i=1}^{n} \frac{\partial Z}{\partial X_i}\left(X_i - \mu_{X_i}\right) + \frac{1}{2!}\sum_{i=1}^{n}\sum_{j=1}^{n} \frac{\partial^2 Z}{\partial X_i \partial X_j}\left(X_i - \mu_{X_i}\right)\left(X_j - \mu_{X_j}\right) \tag{7}$$

$$Z = Z(\mu_X) + \sum_{i=1}^{n} \frac{\partial Z}{\partial X_i}\left(X_i - \mu_{X_i}\right) \tag{7.1}$$

Then the mean and variance are given as:

$$\mu_Z \approx Z\left(\mu_{X_1}, \mu_{X_2}, \dots, \mu_{X_n}\right) \tag{8}$$

$$Var[Z] = Var\left[Z(\mu_X) + \sum_{i=1}^{n} \frac{\partial Z}{\partial X_i}\left(X_i - \mu_{X_i}\right)\right] \tag{9}$$

$$Var[Z] = Var\,[Z(\mu_X)] + Var\left[\sum_{i=1}^{n} \frac{\partial Z}{\partial X_i}(X_i)\right] - Var\left[\sum_{i=1}^{n} \frac{\partial Z}{\partial X_i}\left(\mu_{X_i}\right)\right]_{X_i} \tag{10}$$

$$Var\,[Z(\mu_X)] = 0 \tag{10.1}$$

$$Var\left[\sum_{i=1}^{n} \frac{\partial Z}{\partial X_i}\left(\mu_{X_i}\right)\right]_{X_i} = 0 \tag{10.2}$$

$$Var[Z] = \left[\sum_{i=1}^{n} \frac{\partial Z}{\partial X_i}\right]^2 Var(X_i) \tag{10.3}$$

$$\sigma_Z = \sqrt{\left[\sum_{i=1}^{n} \frac{\partial Z}{\partial X_i}\right]^2 Var(X_i)} \tag{10.4}$$

The reliability index (β) is taken as ratio of the mean to the standard deviation of the performance function.

$$\beta = \frac{\mu_Z}{\sigma_Z} \tag{11}$$

The reliability index is computed for every failure mode, where the probability of failure is expressed as:

$$P_f = \phi(-\beta) = 1 - \phi(\beta) \tag{12}$$

This method is simple to use and assumes that the random variables are normally distributed. All is needed for the calculation is the knowledge of the mean and the standard deviation and it is not necessary to know the distribution of the random variables. The downside of this method is that it can cause error in the final results if the function is nonlinear or if the tail of the distribution cannot be approximated by normal distribution. Moreover, if the function is nonlinear it will be provided different answer than that if it is linear. Advanced FOSM is used to deal with the limitations of the FOSM mentioned above.

3.1.2 Advanced first order second moment method (AFOSM)

AFOSM provides solution for linear and nonlinear performance function by determining the shortest distance from the origin to the failure surface. This method is also called Hasofer-Lind method. It evaluates the probability of failure for the limit state function or the performance function by determining the most probable failure point instead of the mean. Hasofer and Lind developed this advanced method in 1974 which is called Hasofer-Lind method and is abbreviated as H-L method. As stated above, the main objective of this method is to estimate the failure point which is the shortest distance from the origin to the failure surface that separates the failure region from the safe region. This can be clearly shown in **Figure 2**. The failure point is sometimes called in the literature design point or check point, but in this chapter, it will be referred to as the most probable point of failure (MPPF). Let us consider a limit state function/performance function with normally distributed and independent random variables X as:

$$Z(X) = Z(x_1, x_2, \cdots\cdots\cdots\cdots\cdots\cdots, x_n) = 0 \tag{13}$$

This method transforms the random variables into reduced form as:

$$u_i = \frac{x_i - \mu_{x_i}}{\sigma_{x_i}}, \qquad i = 1, 2, \ldots., \tag{14}$$

$$x_i = u_i * \sigma_{x_i} + \mu_{x_i}, \qquad i = 1, 2, \ldots., \tag{15}$$

The performance function is then formulated in terms of the reduced random variables as:

$$Z(U) = Z\left(u_1 * \sigma_{x_1} + \mu_{x_1}, u_2 * \sigma_{x_2} + \mu_{x_2}, \ldots\ldots\ldots\ldots u_n * \sigma_{x_n} + \mu_{x_n}\right) = 0 \tag{16}$$

Figure 2 shows the plot of the limit state function in the original as well as the transformed coordinates. It shows that the MPPF is the tangent point on the curve $Z(X) = 0$ and the reliability index β as the shortest distance from the origin to the limit surface.

To find the MPPF x_i' on the limit surface under the condition that $Z(\mathbf{X}) = 0$, Taylor series expansion is used around the MPPF, considering the first order terms only, this gives:

$$Z(U) = Z(U^*) + \sum_{i=1}^{n} \frac{\partial Z(U^*)}{\partial U_i}(u_i - u_i^*) \tag{17}$$

$$Z(U) = Z(u_1^* * \sigma_{x_1} + \mu_{x_1})(u_2^* * \sigma_{x_2} + \mu_{x_2}) \ldots\ldots\ldots\ldots\ldots\ldots (u_n^* * \sigma_{x_n} + \mu_{x_n}) = 0 \tag{18}$$

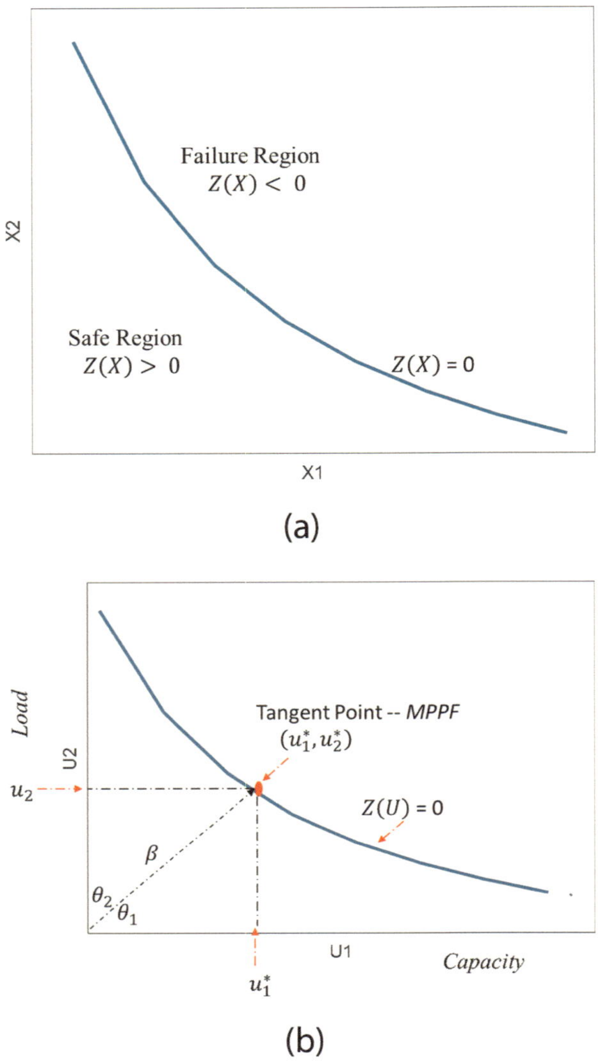

Figure 2.
(a) Nonlinear limit state function—original coordinates. (b) Nonlinear limit state function—transformed coordinates.

Using chain rule for derivative and considering that relationship between U and X as:

$$Z\,function\ of\ x \rightarrow Z = g(x)$$
$$X\,function\ of\ u \rightarrow X = f(u)$$

and using Eqs. (14) and (15) the partial derivative $\frac{\partial g(U^*)}{\partial u_i}$ becomes:

$$\frac{\partial Z}{\partial U_i} = \frac{\partial Z}{\partial X_i}\frac{\partial X}{\partial U_i} = \left(\frac{\partial Z}{\partial X_i}\right)\left(\frac{\partial X\left(u_i^* * \sigma_{x_i} + \mu_{x_i}\right)}{\partial U_i}\right) = \frac{\partial Z}{\partial X_i}\sigma_{x_i} \tag{19}$$

Substituting Eq. (19) into Eq. (17) gives:

$$Z(\boldsymbol{U}) = Z(\boldsymbol{U}^*) + \sum_{i=1}^{n}\frac{\partial Z(\boldsymbol{X}^*)}{\partial x_i}\sigma_{x_i}\left(u_i - u_i^*\right) \tag{20}$$

The mean of $Z(U)$ is:

$$u_z = E[Z(\boldsymbol{U}^*)] + E\left[\sum_{i=1}^{n} \frac{\partial Z(\boldsymbol{X}^*)}{\partial x_i} \sigma_{x_i} u_i\right] - E\left[\sum_{i=1}^{n} \frac{\partial Z(\boldsymbol{X}^*)}{\partial x_i} \sigma_{x_i} u_i^*\right] \tag{21}$$

$$E[Z(U^*)] = Z(U^*) \tag{22}$$

$$E\left[\sum_{i=1}^{n} \frac{\partial Z(\boldsymbol{X}^*)}{\partial x_i} \sigma_{x_i} u_i\right] = E\left[\sum_{i=1}^{n} \frac{\partial Z(\boldsymbol{X}^*)}{\partial x_i} (x_i - u_{x_i})\right] = E[x_i]\left[\sum_{i=1}^{n} \frac{\partial Z(\boldsymbol{X}^*)}{\partial x_i}\right] - E[\mu_{x_i}]\left[\sum_{i=1}^{n} \frac{\partial Z(\boldsymbol{X}^*)}{\partial x_i}\right] \tag{23}$$

$$\mu_{x_i}\left[\sum_{i=1}^{n} \frac{\partial Z(\boldsymbol{X}^*)}{\partial x_i}\right] - \mu_{x_i}\left[\sum_{i=1}^{n} \frac{\partial Z(\boldsymbol{X}^*)}{\partial x_i}\right] = 0 \tag{24}$$

$$E\left[\sum_{i=1}^{n} \frac{\partial Z(\boldsymbol{X}^*)}{\partial x_i} \sigma_{x_i} u_i^*\right] = \sum_{i=1}^{n} \frac{\partial Z(\boldsymbol{X}^*)}{\partial x_i} \sigma_{x_i} u_i^* \tag{25}$$

$$u_z = Z(\boldsymbol{U}^*) - \sum_{i=1}^{n} \frac{\partial Z(\boldsymbol{X}^*)}{\partial x_i} \sigma_{x_i} u_i^* \tag{26}$$

The variance is expressed as:

$$\sigma_Z^2 = Var(Z(\boldsymbol{U})) = Var\,[Z(\boldsymbol{U}^*)] + Var\left[\sum_{i=1}^{n} \frac{\partial Z(\boldsymbol{X}^*)}{\partial x_i} \sigma_{x_i}\,(u_i - u_i^*)\right] \tag{27}$$

It must be noted that constants have no variance, their variances equal to zero. The first term of the Taylor expansion in Eq. (27) is constant; therefore, its variance equals to zero. Similarly, the variance at the mean value is zero.

$$Var\,[Z(\boldsymbol{U}^*)] = 0 \tag{28}$$

$$Var(Z(\boldsymbol{U})) = Var\left[\sum_{i=1}^{n} \frac{\partial Z(\boldsymbol{X}^*)}{\partial x_i} \sigma_{x_i}\,(u_i)\right] - Var\left[\sum_{i=1}^{n} \frac{\partial Z(\boldsymbol{X}^*)}{\partial x_i} \sigma_{x_i}\,(u_i^*)\right] \tag{29}$$

$$Var\left[\sum_{i=1}^{n} \frac{\partial Z(\boldsymbol{X}^*)}{\partial x_i} \sigma_{x_i}\,(u_i^*)\right] = 0 \tag{30}$$

$$Var(Z(\boldsymbol{U})) = Var\left[\sum_{i=1}^{n} \frac{\partial Z(\boldsymbol{X}^*)}{\partial x_i} \sigma_{x_i}\,(u_i)\right] = Var\left[\sum_{i=1}^{n} \frac{\partial Z(\boldsymbol{X}^*)}{\partial x_i}\,(x_i - u_{x_i})\right] \tag{31}$$

$$Var\left[\sum_{i=1}^{n} \frac{\partial Z(\boldsymbol{X}^*)}{\partial x_i}\,(x_i - u_{x_i})\right] = Var\left[\sum_{i=1}^{n} \frac{\partial Z(\boldsymbol{X}^*)}{\partial x_i} x_i\right] - Var\left[\sum_{i=1}^{n} \frac{\partial Z(\boldsymbol{X}^*)}{\partial x_i} u_{x_i}\right] \tag{31.1}$$

$$Var\left[\sum_{i=1}^{n} \frac{\partial Z(\boldsymbol{X}^*)}{\partial x_i} u_{x_i}\right] = \mathbf{0} \tag{32}$$

$$Var(Z(\boldsymbol{U})) = \sum_{i=1}^{n} \left(\frac{\partial Z(\boldsymbol{X}^*)}{\partial x_i}\right)^2 Var\,[x_i] \tag{33}$$

$$\sigma_z = \sqrt{\sum_{i=1}^{n} \left(\frac{\partial Z(\boldsymbol{X}^*)}{\partial x_i}\right)^2 \sigma_{x_i}^2} \tag{34}$$

Alternatively, Eq. (31) can be calculated as [6]:

$$Var\left[\sum_{i=1}^{n}\frac{\partial Z(\boldsymbol{X}^*)}{\partial x_i}\left(x_i-u_{x_i}\right)\right]=\boldsymbol{E}\left[\sum_{i=1}^{n}\frac{\partial Z(\boldsymbol{X}^*)}{\partial x_i}\left(x_i-u_{x_i}\right)\right]^2$$
$$=\sum_{i=1}^{n}\left(\frac{\partial Z(\boldsymbol{X}^*)}{\partial x_i}\right)^2 E[(x_i-u_{x_i})]^2 \quad (34.1)$$
$$=\sum_{i=1}^{n}\left(\frac{\partial Z(\boldsymbol{X}^*)}{\partial x_i}\right)^2 Var\left[x_i\right]$$

The reliability index is calculated as:

$$\beta=\frac{\mu_Z}{\sigma_Z}=\frac{Z(\boldsymbol{X}^*)-\sum_{i=1}^{n}\frac{\partial Z(\boldsymbol{X}^*)}{\partial x_i}\sigma_{x_i}\left(u_i^*\right)}{\sqrt{\sum_{i=1}^{n}\left(\frac{\partial Z(\boldsymbol{X}^*)}{\partial x_i}\sigma_{x_i}\right)^2}} \quad (35)$$

The directional cosine α_i along the coordinate axes is computed as:

$$\alpha_i=-\frac{\left(\frac{\partial Z(\boldsymbol{U}^*)}{\partial u_i}\right)}{\sqrt{\sum_{i=1}^{n}\left(\frac{\partial Z(\boldsymbol{U}^*)}{\partial u_i}\right)^2}}=-\frac{\left(\frac{\partial Z(\boldsymbol{X}^*)}{\partial x_i}\sigma_{x_i}\right)}{\sqrt{\sum_{i=1}^{n}\left(\frac{\partial Z(\boldsymbol{X})}{\partial x_i}\sigma_{x_i}\right)^2}} \quad (36)$$

It can be shown from **Figure 2** that:

$$u_i^*=\beta * Cos\left(\theta_i\right) \; or \; u_i^*=\alpha_{x_i}\beta \quad (37)$$

Using Eqs. (14), (15) and (36) we can determine the design point in the original coordinates as:

$$x_i^*=\beta\alpha_{x_i}\sigma_{x_i}+\mu_{x_i}=u_i^*\sigma_{x_i}+\mu_{x_i} \; i=1,2,\ldots\ldots\ldots\ldots\ldots,n \quad (38)$$

The probability of failure, P_f can be computed as:

$$P_f=\Phi(-\beta)=1-\Phi(\beta) \quad (39)$$

The steps to estimate the reliability index are:

1. *Formulate the performance function in terms of the original random variables,* x_i*:*

$$Z(x_1,x_2,\cdots\cdots\cdots\cdots\cdots\cdots,x_n)$$

2. *Assume the initial design points as the given mean of each variables.*

3. *Compute the initial reliability index* β *in terms of the mean values of the random variables using Eq. (11).*

4. *Compute the partial derivatives of the limit state function (LSF)/performance function in terms of mean value of the random variables.*

5. *Compute the directional cosines* α_i *using Eq. (36).*

6. *Compute the new design point using Eq. (38).*

7. *Compute the LSF in terms of new design points.*

8. *Compute the partial derivatives at the new design points.*

9. *Compute the new reliability index β using Eq. (35).*

   a. *Another alternative is to compute β from the limit state function $\left(Z(X) = Z\left(\beta\alpha_{x_i}\sigma_{x_i} + \mu_{x_i}\right) = 0\right)$ using the newly determined design points in step 7 and solve for β.*

10. *Compute the directional cosine.*

11. *Repeat steps six through nine until β converges to a pre-established tolerance level.*

12. *Use Eq. (39) to calculate the probability of failure.*

The steps mentioned above are used with the assumption that the random variables are normally distributed. For non-normally distributed variables additional steps are needed to determine the mean and standard deviation of the equivalent normal distribution as listed below. These steps should be carried out after step number two to determine mean and standard deviation of the equivalent normal distribution. Assuming the random variables are statistically independent and non-normally distributed:

1. Determine the distribution parameters.

2. Compute the cumulative distribution function *cdf*, $F(x_i)$, the probability density function *pdf*, $f(x_i)$ and the inverse *cdf*, $\Phi^{-1}[F(x_i)]$ of the original non-normal random variables at the initial design point.

3. Compute the values of the standard deviation, σ_{x_i} and the mean, μ_{x_i} of the equivalent normal distribution as:

$$\sigma_{x_i} = \frac{\phi\left(\Phi^{-1}[F(x_i)]\right)}{f(x_i)} \tag{40}$$

It must be noted that $f(x_i)$ refers to the pdf of the original non-normal random variable and $\phi(\ )$ refers to the pdf of the equivalent standard normal random variable.

$$\mu_{x_i} = x_i - \Phi^{-1}[\,F(x_i)\,]\sigma_{x_i} \tag{41}$$

4. Compute the standard normal variable (u_i) of x_i by:

$$u_i = \frac{x_i - \mu_{x_i}}{\sigma_{x_i}}, i = 1, 2, \dots\dots\dots\dots\dots\dots, n \tag{42}$$

For a log normally distributed random variable, distribution parameters $\mu_{\ln(x)}$, $\sigma_{\ln(x)}$ are defined using the following equations:

$$\sigma^2_{\ln(x)} = \ln\left(\left(\frac{\sigma_x}{\mu_x}\right)^2 + 1\right), \qquad \sigma_{\ln(x)} = \sqrt{\sigma^2_{\ln(x)}} \tag{43}$$

$$\mu_{\ln(x)} = \ln(\mu_x) - 0.5\sigma^2_{\ln(x)} \tag{44}$$

The pdf and *cdf* are defined as:

$$f(x) = \frac{1}{\sqrt{2\pi}\,(x)\sigma_{\ln(x)}}\ \exp\left(-0.5\left(\frac{\ln(x) - \mu_{\ln(x)}}{\sigma_{\ln(x)}}\right)^2\right) \tag{45}$$

$$F(x) = \Phi\left(\frac{\ln(x) - \mu_{\ln(x)}}{\sigma_{\ln(x)}}\right) \tag{46}$$

For other distribution types the readers are referred to Refs. [4, 7, 8].

3.2 Simulation methods

Alternatively, the probability of failure is computed using Monte Carlo simulation method. Two methods are considered in this chapter, the counting and sample statistics methods. The simulation is conducted using computer programs such MATLAB, C++ or MINITAB or any other simulation programming packages.

3.2.1 Monte Carlo counting method

The counting method is formulated by dividing the number of simulation cycles at the events when the *Z* function becomes less than 0 ($N_f < 0$) by the total number of simulation cycles (N).

$$P_f = \frac{N_f}{N} \tag{47}$$

The steps for Monte Carlo simulation counting method are listed below:
Counting method

1. *Formulate the performance function in terms of the original random variables, x_i:*

$$Z(x_1, x_2, \cdots\cdots\cdots\cdots\cdots\cdots, x_n)$$

2. *Determine the distribution and its parameters for each random variable.*

3. *Assign N to be the number of simulation cycles.*

4. *Assign M to be the number of calculation times.*

5. *Assign Nf to be the number of simulation cycles when the Z function becomes less than 0.*

6. *Initialize:*

    a. *Set M to zero*

    b. *Set Nf to zero*

7. *Generate random values from the given distribution. With the determined distribution parameters for each variable.*

    a. *As an example, if the random variable is normally distributed use the following Matlab function to generate the random values:*

$$x = normrnd(mu, \text{var}, n, 1) \tag{48}$$

N is number of simulation cycles, mu and var. are the distribution parameters (the mean and standard deviation of the random variable).

b. *If the random variable is log normally distributed with a mean, μ_{x_i} and standard deviation, σ_{x_i}:*

i. *Determine the distribution parameters; Eqs. (43) and (44) can be used use to calculate these two parameters.*

ii. *Use the following Matlab function to generate the random values:*

$$x = lognrnd\left(\mu_{\ln(x)}, \sigma_{\ln(x)}, N, 1\right) \tag{49}$$

8. *Calculate Pf:*

a. *Is Z < 0?*

i. *If yes:*

- *Nf = Nf +1,*
- *M = M + 1,*

ii. *If no:*

- *Is M = N?*

a. *If no—go to step 7*

b. *If yes—calculate Pf* $= \frac{Nf}{N}$

Alternative approach to random number generation is to use the following steps:

- *Generate uniformly distributed random numbers u_i between 0 and 1; this can be accomplished by using software packages such as MATLAB, excel or C++ and other software packages.*
- *Equate the inverse cumulative distribution function cdf of the random variable to the generated random numbers, u_i by using the following equation:*

$$u_i = F_x(x_i) \tag{49.1}$$

$$x_i = F_x^{-1}(u_i) \tag{49.2}$$

As an example, if x_i follows normal distribution with a mean μ_{x_i}, and standard deviation σ_{x_i} then its random number becomes:

$$x_i = \mu_{x_i} + \sigma_{x_i}\left(F_x^{-1}(u_i)\right) \tag{49.3}$$

$\left(F_x^{-1}(u_i)\right)$ *can be determined using the cdf of the standard normal distribution, these tables are included in most probability and statistics books.*

The random numbers can be generated N times using the following MATALB function:

$$u_i = \text{rand}(N, 1) \tag{49.4}$$

3.2.2 Monte Carlo sample statistics method

The Monte Carlo sample statistics methods considers the mean μ_z and the standard deviation σ_z in computing the reliability index.

$$\beta = \frac{\mu_z}{\sigma_z} \tag{50}$$

$$P_f = \Phi(-\beta) = 1 - \Phi(\beta) \tag{51}$$

The steps for Monte Carlo simulation sample statistics method are listed below: Sampling method

1. *Formulate the performance function in terms of the original random variables, x_i: $Z(x_1, x_2, \cdots \cdots\cdots\cdots\cdots\cdots, x_n)$.*

2. *Determine the distribution and its parameters for each random variable.*

3. *Assign N to be the number of simulation times.*

4. *Assign M to be the number of calculation times.*

5. *Initialize:*

    a. *Set M to zero*

    b. *Set Nf to zero*

6. *Generate random values from the given distribution with the determined distribution parameters for each variable.*

7. *Calculate Z function.*

8. *Calculate Pf:*

    c. *Calculate* $\beta = \frac{\mu_z}{\sigma_z}$

    d. *Calculate* $P_f = 1 - \Phi(\beta)$

    e. *Is M = N?*

    i. *If no—go to step 6*

    ii. *If yes—stop*

It must be noted that Monte Carlo sample statistics method can be used only for linear functions having uncorrelated normal random variables.

Example 1

The performance function for a system has been formulated as: $z = 2.5x_1 - x_2$ *where*

$$x_1 \sim N(2, 0.2), x_2 \sim N(3.1, 0.32)$$

a. Determine the probability of failure using simulation methods: Monte Carlo (MCS) sample statistics and Monte Carlo counting simulation methods.

b. Determine the probability of failure using the analytical methods: FOSM and AFOSM (Hasofer-Lind) methods.

Solution:

A. Monte Carlo simulation was conducted for both methods, the sample statistics and counting methods. The number of simulation cycles used in the analysis are 2e5 and 1e6 cycles (**Table 1**).

B. See **Table 2**.

Example 2

The performance function for a leak detection system has been formulated as [2]: $z = 2.8252x^{-1} - x_2$, $x_1 \sim N(1.1, 0.02), x_2 \sim N(2.37, 0.04)$

a. Determine the probability of failure using Monte Carlo counting simulation method.

b. Determine the probability of failure using AFOSM method.

This example is adopted from reference [1, 2], with some modifications.

Solution

Part a

Monte Carlo simulation was conducted for both methods, the sample statistics and counting methods as indicated in the table.

Table 3 indicates that the probability of failure converges to *0.000476*.

	MCS—sample statistics method		*MCS—counting method*	
Number of simulation cycles	***Reliability index—β***	***Pf***	***Reliability index—β***	***Pf***
2e5	*3.1993*	*0.000689*	*3.1967*	*0.000695*
1e6	*3.2030*	*0.00068*	*3.2005*	*0.000686*

Table 1.
Simulation methods results—Example 1.

FOSM		*AFOSM*	
Reliability index—β	***Pf***	***Reliability index—β***	***Pf***
3.2006	*0.000686*	*3.2006*	*0.000686*

***This example adopted from [9].*

Table 2.
Analytical methods results—Example 1.

	MCS—counting method	
Number of simulation cycles	Reliability index—β	Pf
1e5	3.2962	0.000490
1e6	3.3043	0.000476

Table 3.
Simulation methods results—Example 2.

Part b

First the Z function is computed in terms of the mean values of the random variables:

$$x_1 = \mu_{x_1} = 1.1$$

$$x_2 = \mu_{x_2} = 2.37$$

$$u_1 = \frac{x_1 - \mu_{x_1}}{\sigma_{x1}} = \frac{1.1 - 1.1}{0.02} = 0$$

$$u_2 = \frac{x_2 - \mu_{x_2}}{\sigma_{x2}} = \frac{2.37 - 2.37}{0.04} = 0$$

$$Z(X) = Z(1.1, 2.37) = 0.1984$$

The partial derivatives:

$$\frac{\partial z}{\partial x_1} = \frac{-2.8252}{{x_1}^2} = -2.3349$$

$$\frac{\partial z}{\partial x_2} = -1$$

$$\sigma_z = \sqrt{\sum_{i=1}^{n}\left(\frac{\partial Z(X)}{\partial x_i}\sigma_{x_i}\right)^2} = \sqrt{(\mathbf{-2.3349\ \ 0.02})^2 + (\mathbf{-1\ \ 0.04})^2} = 0.061487$$

$$\alpha_1 = -\frac{\left(\frac{\partial Z(X)}{\partial x_1}\sigma_{x_1}\right)}{\sqrt{\sum_{i=1}^{n}\left(\frac{\partial Z(X)}{\partial x_i}\sigma_{xi}\right)^2}} = -\frac{(-2.3349 * 0.02)}{0.061487} = 0.75947$$

$$\alpha_2 = -\frac{\left(\frac{\partial Z(X)}{\partial x_2}\sigma_{x_2}\right)}{\sqrt{\sum_{i=1}^{n}\left(\frac{\partial g(X)}{\partial x_i}\sigma_{xi}\right)^2}} = -\frac{(-1 * 0.04)}{0.061487} = 0.65054$$

$$\beta = \frac{\mu_z}{\sigma_z} = \frac{0.19836}{0.061487} = 3.2261$$

$$P_f = \Phi(-\beta) = 1 - \Phi(\beta) = 1 - \Phi(3.2261) = 0.000627$$

Iteration 1

$$u_1 = \beta\alpha_{x_1} = 3.2261 * 0.75947 = 2.45013$$

$$u_2 = \beta\alpha_{x_2} = 3.2261 * 0.65054 = 2.09872$$

$$x_1 = \beta\alpha_1\sigma_{x_1} + \mu_{x_1} = u_1\sigma_{x_1} + \mu_{x_1} = 2.45013 * 0.02 + 1.1 = 1.149003$$

$$x_2 = \beta\alpha_2\sigma_{x_2} + \mu_{x_2} = u_2\sigma_{x_2} + \mu_{x_2} = 2.09872 * 0.04 + 2.37 = 2.453949$$

$$Z(X) = Z\ (1.149003, 2.453949) = 0.004880$$

$$\frac{\partial Z}{\partial x_1} = \frac{-2.8252}{{x_1}^2} = \frac{-2.8252}{(1.149003)^2} = -2.13997$$

$$\frac{\partial Z}{\partial x_2} = -1$$

$$\alpha_1 = -\frac{\left(\frac{\partial Z(X)}{\partial x_1}\sigma_{x_1}\right)}{\sqrt{\sum_{i=1}^{n}\left(\frac{\partial Z(X)}{\partial x_i}\sigma_{xi}\right)^2}} = -\frac{(-2.13997 * 0.02)}{\sqrt{(-2.13997 * 0.02)^2 + (-1 * 0.04)^2}} = \frac{0.042799}{0.05858} = 0.7306$$

$$\alpha_2 = -\frac{\left(\frac{\partial Z(X)}{\partial x_2}\sigma_{x_2}\right)}{\sqrt{\sum_{i=1}^{n}\left(\frac{\partial Z(X)}{\partial x_i}\sigma_{xi}\right)^2}} = -\frac{(-1 * 0.04)}{\sqrt{(-2.13997 * 0.02)^2 + (-1 * 0.04)^2}} = 0.6828$$

$$\beta_{HL} = \frac{Z(X) - \sum_{i=1}^{n}\frac{\partial Z(X)}{\partial x_i}\sigma_{x_i}(u_i)}{\sqrt{\sum_{i=1}^{n}\left(\frac{\partial Z}{\partial x_i}\sigma_{x_i}\right)^2}}$$

$$= \frac{0.004880 - ((-2.13997 * 0.02 * 2.45013) + (-1 * 0.04 * 2.09872)}{0.05858}$$

$$= \frac{0.19369}{0.05858} = 3.3064$$

$$P_f = \Phi(-\beta) = 1 - \Phi(\beta) = 1 - \Phi(3.3064) = 0.0004726$$

The probability of failure obtained by counting method is very close to that obtained by AFOSM (**Table 4**).

Example 3

A pipeline segment is suffering corrosion that grows annually at steady rate. The extent of the initial growth has been estimated to be 4.7 mm and it is assumed that it follows log normal distribution with standard deviation of 1.1.

The corrosion annual growth follows log normal distribution with a mean and standard deviation values of 0.2 and 0.01. The pipeline wall thickness follows normal distribution with a mean and standard deviation values of 14 mm and 4.7 respectively. The critical pipeline wall thickness has been determined to be 80% of the wall thickness [3]. A summary of the relevant information pertaining to the pipeline corrosion is presented in the net table, **Table 5**.

The owner of the pipeline decides not to repair the corrosion and wants to know if the pipeline can survive for the next 14 years without causing a leakage. It has been decided that in order to be in the safe side the maximum acceptable probability of failure has been set to 1e4 [3].

Solve this problem using analytical method as well as Monte Carlo (MCS) simulation method.

Solution

Monte Carlo simulation

x_1: Wall thickness, $x_1 \sim N(1.1, 0.02)$

x_2: Extent of the initial corrosion, $x_2 \sim LGN(4.7, 1.1)$

x_3: Corrosion annual growth, $x_3 \sim LGN(0.2, 0.01)$

The limit state function can be formulated as indicated below:

The capacity portion is formulated as: $0.8x_1$

The demand portion is formulated as: $x_2 + 14x_3$

Iteration	x1	x2	u1	u2	Z(X)	∂z/∂x1	∂z/∂x2	α1	α2	β	pf
1	1.100000	2.370000	0.000000	0.000000	0.198364	−2.334876	−1	0.759469	0.650543	3.226104	0.000627
2	1.149003	2.453949	2.450126	2.098721	0.004880	−2.139968	−1	0.730596	0.682810	3.306375	0.000473
3	1.148312	2.460305	2.415624	2.257626	0.000001	−2.142540	−1	0.731005	0.682372	3.306389	0.000473
4	1.148340	2.460248	2.416987	2.256188	0.000000	−2.142439	−1	0.730989	0.682389	3.306389	0.000473

Table 4.
Summary of the analytical methods results—Example 2.

	Wall thickness	*Initial corrosion*	*Corrosion growth*
Distribution type	*Normal distribution*	*Log normal distribution*	*Log normal distribution*
Mean	*14 mm*	*4.7 mm*	*0.2 mm*
Standard deviation	*0.07*	*1.1*	*0.01*

Table 5.
Pipeline corrosion data—Example 3.

$$Z(X) = 0.8x_1 - (x_2 + 14x_3)$$

Pipeline wall thickness:

$$\mu_{x_1} = 14, \sigma_{x_1} = 0.07$$

$$x_1 = normrnd(\mu_{x_1}, \sigma_{x_1}, N, 1)$$

Initial corrosion:

$$\mu_{x_2} = 4.7, \sigma_{x_2} = 1.1$$

$$\sigma^2_{\ln(x_2)} = \ln\left(\left(\frac{\sigma_{x_2}}{\mu_{x_2}}\right)^2 + 1\right) = 0.0533$$

$$\sigma_{\ln(x_2)} = \sqrt{\sigma^2_{\ln(x_2)}} = 0.2309$$

$$\mu_{\ln(x_2)} = \ln(\mu_{x_2}) - 0.5\sigma^2_{\ln(x_2)} = 1.5209$$

$$x_2 = lognrnd\left(\mu_{\ln(x_2)}, \sigma_{\ln(x_2)}, N, 1\right)$$

Corrosion growth rate:

$$\mu_{x_3} = 0.2, \sigma_{x_3} = 0.01$$

$$\sigma^2_{\ln(x_3)} = \ln\left(\left(\frac{\sigma_{x_3}}{\mu_{x_3}}\right)^2 + 1\right) = 0.002497$$

$$\sigma_{\ln(x_3)} = \sqrt{\sigma^2_{\ln(x_3)}} = 0.04997$$

$$\mu_{\ln(3)} = \ln(\mu_{x_3}) - 0.5\sigma^2_{\ln(x_3)} = -1.61069$$

$$x_3 = lognrnd\left(\mu_{\ln(x_3)}, \sigma_{\ln(x_3)}, N, 1\right)$$

Monte Carlo simulation produced the following results as shown in **Table 6**. The probability of failure and beta converge to the following values:

$$Pf = \frac{Nf}{N} = \frac{4396}{1e6} = 0.004389$$

$$Beta = \Phi^{-1}(Pf) = 2.6206$$

	MCS—counting method	
Number of simulation cycles	*Reliability index—β*	*Pf*
1e6	2.6249	0.004334
1e7	2.6206	0.004389

Table 6.
Simulation methods results—Example 3.

Here only the counting method is used because the sampling method produces different results. The sampling method produces accurate results for linear and normal limit state/performance function only.

Analytical Solution

Assume the initial value for each random variable to be its mean.

$$Z(X) = Z\ (14, 4.7, 0.2) = 3.7$$

$$\frac{\partial Z}{\partial x_1} = 0.8$$

$$\frac{\partial Z}{\partial x_2} = -1$$

$$\frac{\partial Z}{\partial x_3} = -14$$

For non-normal variables, the standard deviation and mean values of the equivalent normal variables are calculated using Eqs. (40) and (41).

Pipeline wall thickness, x_1:

$$\mu_{x_1} = 14, \sigma_{x_1} = 0.07$$

Initial corrosion, x_2:

$$\mu_{x_2} = 4.7, \sigma_{x_2} = 1.1$$

$$\sigma^2_{\ln(x_2)} = \ln\left(\left(\frac{\sigma_{x_2}}{\mu_{x_2}}\right)^2 + 1\right) = 0.0533$$

$$\sigma_{\ln(x_2)} = \sqrt{\sigma^2_{\ln(x_2)}} = 0.2309$$

$$\mu_{\ln(x_2)} = \ln\left(\mu_{x_2}\right) - 0.5\sigma^2_{\ln(x_2)} = 1.5209$$

Compute the pdf of the original non-normal variable (log normal distribution) using Eq. (45):

$$f(x) = \frac{1}{\sqrt{2\pi}\ (x)\sigma_{\ln(x)}} \exp\left(-0.5\left(\frac{\ln(x) - \mu_{\ln(x)}}{\sigma_{\ln(x)}}\right)^2\right)$$

$$= \frac{1}{\sqrt{2\pi}(4.7)(0.2309)} \exp\left(-0.5\left(\frac{\ln(4.7) - 1.5209}{0.2309}\right)^2\right) = 0.36512$$

Compute the *cdf* of the original non-normal variable (log normal distribution) using Eq. (46):

$$F(x) = \Phi\left(\frac{\ln(x_2) - \mu_{\ln(x)}}{\sigma_{\ln(x)}}\right) = \Phi\left(\frac{\ln(4.7) - 1.5209}{0.2309}\right)$$

$$\Phi^{-1}[F(x_i)] = \left(\frac{\ln(4.7) - 1.5209}{0.2309}\right) = 0.115466$$

Compute the mean and standard deviation of the equivalent normal variable at the design point using Eqs. (40) and (41):

$$\sigma_{x_2} = \frac{\phi(\Phi^{-1}[F(x_2)])}{f(x_2)} = \frac{\phi\,(0.115466)}{0.36512},$$

$\phi(\,)$ refers to the equivalent standard normal random variable and is calculated as:

$$\phi\,(0.115466) = \frac{1}{\sqrt{2\pi}}\ \exp\left(-0.5\left(\frac{\ln(x_2) - \mu_{\ln(x)}}{\sigma_{\ln(x)}}\right)^2\right)$$

$$= \frac{1}{\sqrt{2\pi}}\ \exp\left(-0.5(0.115466)^2\right) = 0.3963$$

$$\sigma_{x_2} = \frac{0.3963}{0.36512} = 1.0854$$

$$\mu_{x_2} = x_2 - \Phi^{-1}[F(x_2)]\sigma_{x_2} = 4.7 - 0.115466(1.0854) = 4.5747$$

Corrosion growth, x_3

The same procedures outline for initial corrosion extent, x_2: to produces the following results:

$$f(x_3) = 39.907,\ \Phi^{-1}[F(x_3)] = 0.02526,\ \phi(\Phi^{-1}[F(x_3)]) = 0.398815,$$

$$\sigma_{x_3} = 0.0099936\ and\ \mu_{x_3} = 0.19975$$

$$\sigma_z = \sqrt{\sum_{i=1}^{n}\left(\frac{\partial Z(X)}{\partial x_i}\sigma_{x_i}\right)^2} = 1.0958$$

$$\beta = \frac{\mu_z}{\sigma_z} = \frac{3.7}{1.0958} = 3.3766$$

$$P_f = \Phi(-\beta) = 0.00037$$

$$\alpha_1 = -\frac{\left(\frac{\partial Z(X)}{\partial x_1}\sigma_{x_1}\right)}{\sqrt{\sum_{i=1}^{n}\left(\frac{\partial Z(X)}{\partial x_1}\sigma_{x1}\right)^2}} = -0.0511$$

$$\alpha_2 = -\frac{\left(\frac{\partial Z(X)}{\partial x_2}\sigma_{x_2}\right)}{\sqrt{\sum_{i=1}^{n}\left(\frac{\partial Z(X)}{\partial x_2}\sigma_{x2}\right)^2}} = 0.9905$$

$$\alpha_3 = -\frac{\left(\frac{\partial Z(X)}{\partial x_3}\sigma_{x_3}\right)}{\sqrt{\sum_{i=1}^{n}\left(\frac{\partial Z(X)}{\partial x_3}\sigma_{x3}\right)^2}} = 0.1277$$

Iteration 1

$$u_1 = \beta\alpha_{x_1} = -0.1013$$

$$u_2 = \beta\alpha_{x_2} = 3.3445$$

$$u_3 = \beta\alpha_{x_3} = 0.4311$$

$$x_1 = \beta\alpha_1\sigma_{x_1} + \mu_{x_1} = u_1\sigma_{x_1} + \mu_{x_1} = 13.9879$$

$$x_2 = \beta\alpha_2\sigma_{x_2} + \mu_{x_2} = u_2\sigma_{x_2} + \mu_{x_2} = 8.2047$$

$$x_3 = \beta\alpha_3\sigma_{x_3} + \mu_{x_3} = u_2\sigma_{x_3} + \mu_{x_3} = 0.20406$$

$$\frac{\partial g}{\partial x_1} = 0.8, \quad \frac{\partial g}{\partial x_2} = -1, \quad \frac{\partial g}{\partial x_3} = -14$$

Iteration	1	2	3	4	5	6
$x1$	14.0000	13.9879	13.9929	13.9953	13.9947	13.9947
$x2$	4.7000	8.2047	9.9091	8.5641	8.4595	8.3734
$x3$	0.2000	0.2041	0.2023	0.2015	0.2017	0.2016
$u1$	0.0000	−0.1726	−0.1013	−0.0675	−0.0753	−0.0750
$u2$	0.0000	3.3445	3.4276	2.7576	2.6609	2.6164
$u3$	0.0000	0.4311	0.2582	0.1706	0.1896	0.1889
$mu1$	14.0000	14.0000	14.0000	14.0000	14.0000	14.0000
$mu2$	4.5747	3.4147	2.2538	3.1972	3.2620	3.3145
$mu3$	0.1998	0.1997	0.1997	0.1997	0.1997	0.1997
$\sigma 1$	0.0700	0.0700	0.0700	0.0700	0.0700	0.0700
$\sigma 2$	1.0854	1.8947	2.2883	1.9777	1.9536	1.9337
$\sigma 3$	0.0100	0.0102	0.0101	0.0101	0.0101	0.0101
$Z(x)$	3.7000	0.1288	−1.5475	−0.1882	−0.0869	−0.0006
$\partial z/\partial x1$	0.8000	0.8000	0.8000	0.8000	0.8000	0.8000
$\partial z/\partial x2$	−1.0000	−1.0000	−1.0000	−1.0000	−1.0000	−1.0000
$\partial z/\partial x3$	−14.0000	−14.0000	−14.0000	−14.0000	−14.0000	−14.0000
σ	1.0958	1.9009	2.2933	1.9835	1.9594	1.9396
$\alpha 1$	−0.0511	−0.0295	−0.0244	−0.0282	−0.0286	−0.0289
$\alpha 2$	0.9905	0.9967	0.9978	0.9971	0.9970	0.9969
$\alpha 3$	0.1277	0.0751	0.0617	0.0711	0.0720	0.0727
β	3.3766	3.4388	2.7637	2.6687	2.6243	2.6240
Pf	0.0004	0.0003	0.0029	0.0038	0.0043	0.0043

Table 7.
Analytical methods results—Example 3.

Analytical method		MCS—counting method	
Reliability index—β	Pf	Reliability index—β	Pf
2.6240	0.004345	2.6206	0.004389

Table 8.
Comparison of results—Example 3.

System A	Distribution type and its parameters	
Electric motor	Capacity	Demand
System energy	N~(0.52, 0.05)	N~(0.46, 0.08)
Number of parts	N~(2, 0.5)	N~(1, 0.3)
Weight	N~(0.42 0.05)	N~(0.400, 0.06)
System B	**Distribution type and its parameters**	
Pneumatic cylinder	**Capacity**	**Demand**
System energy	N~(0.52, 0.05)	N~(0.4, 0.04)
Number of parts	N~(2, 0.5)	N~(1, 0.3)
Weight	N~(0.420, 0.05)	N~(0.37, 0.04)

Table 9.
Capacity and demand variables—Example 4.

System A	MCS—sample statistics method		MCS—counting method	
Number of simulation cycles	Reliability index—β	Pf	Reliability index—β	Pf
Electric motor—system energy				
1e5	0.6334	0.2632	0.6326	0.2635
1e6	0.6363	0.2623	0.6363	0.2623
Electric motor—number of parts				
1e5	1.7207	0.0427	1.7197	0.0427
1e6	1.7168	0.0430	1.7183	0.0429
Electric motor—weight				
1e5	0.2556	0.3991	0.2572	0.3985
1e6	0.2558	0.3991	0.2564	0.3988
System B	**MCS—sample statistics method**		**MCS—counting method**	
Number of simulation cycles	**Reliability index—β**	**Pf**	**Reliability index—β**	**Pf**
	Pneumatic cylinder—system energy			
1e5	1.8808	0.0300	1.8927	0.0292
1e6	1.8748	0.0304	1.8770	0.0303
	Pneumatic cylinder—number of parts			
1e5	1.7088	0.0437	1.7036	0.0442
1e6	1.7150	0.0432	1.7156	0.0431
	Pneumatic cylinder—weight			
1e5	0.7808	0.2175	0.7801	0.2177
1e6	0.781443467	0.2172709	0.78127631	0.21732

Table 10.
Simulation methods results systems A and B—Example 4.

							Electric motor—system energy							
Iteration	***x1***	***x2***	***u1***	***u2***	***σ1***	***σ2***	***Z(X)***	***∂z/∂x1***	***∂z/∂x2***	***α1***	***α2***	***σ_z***	***β***	***pf***
1	0.5200	0.4600	0.0000	0.0000	0.0500	0.0800	0.0600	1.0000	−1.0000	−0.5300	0.8480	0.0943	0.6360	0.2624
2	0.5031	0.5031	−0.3371	0.5393	0.0500	0.0800	0.0000	1.0000	−1.0000	−0.5300	0.8480	0.0943	0.6360	0.2624
							Electric motor—number of parts							
Iteration	***x1***	***x2***	***u1***	***u2***	***σ1***	***σ2***	***Z(X)***	***∂z/∂x1***	***∂z/∂x2***	***α1***	***α2***	***σ_z***	***β***	***pf***
1	2.0000	1.0000	0.0000	0.0000	0.5000	0.3000	1.0000	1.0000	−1.0000	−0.8575	0.5145	0.5831	1.7150	0.0432
2	1.2647	1.2647	−1.4706	0.8824	0.5000	0.3000	0.0000	1.0000	−1.0000	−0.8575	0.5145	0.5831	1.7150	0.0432
							Electric motor—weight							
Iteration	***x1***	***x2***	***u1***	***u2***	***σ1***	***σ2***	***Z(X)***	***∂z/∂x1***	***∂z/∂x2***	***α1***	***α2***	***σ_z***	***β***	***pf***
1	0.4200	0.4000	0.0000	0.0000	0.0500	0.0600	0.0200	1.0000	−1.0000	−0.6402	0.7682	0.0781	0.2561	0.3989
2	0.4118	0.4118	−0.1639	0.1967	0.0500	0.0600	0.0000	1.0000	−1.0000	−0.6402	0.7682	0.0781	0.2561	0.3989
							Pneumatic cylinder—system energy							
Iteration	***x1***	***x2***	***u1***	***u2***	***σ1***	***σ2***	***Z(X)***	***∂z/∂x1***	***∂z/∂x2***	***α1***	***α2***	***σ_z***	***β***	***pf***
1	0.5200	0.4000	0.0000	0.0000	0.0500	0.0400	0.1200	1.0000	−1.0000	−0.7809	0.6247	0.0640	1.8741	0.0305
2	0.4468	0.4468	−1.4634	1.1707	0.0500	0.0400	0.0000	1.0000	−1.0000	−0.7809	0.6247	0.0640	1.8741	0.0305
							Pneumatic cylinder—number of parts							
Iteration	***x1***	***x2***	***u1***	***u2***	***σ1***	***σ2***	***Z(X)***	***∂z/∂x1***	***∂z/∂x2***	***α1***	***α2***	***σ_z***	***β***	***pf***
1	2.0000	1.0000	0.0000	0.0000	0.5000	0.3000	1.0000	1.0000	−1.0000	−0.8575	0.5145	0.5831	1.7150	0.0432
2	1.2647	1.2647	−1.4706	0.8824	0.5000	0.3000	0.0000	1.0000	−1.0000	−0.8575	0.5145	0.5831	1.7150	0.0432
							Pneumatic cylinder—weight							
Iteration	***x1***	***x2***	***u1***	***u?***	***σ1***	***σ2***	***Z(X)***	***∂z/∂x1***	***∂z/∂x2***	***u1***	***α2***	***σ_z***	***β***	***pf***
1	0.4200	0.3700	0.0000	0.0000	0.0500	0.0400	0.0500	1.0000	−1.0000	−0.7809	0.6247	0.0640	0.7809	0.2174
2	0.3895	0.3895	−0.6098	0.4878	0.0500	0.0400	0.0000	1.0000	−1.0000	−0.7809	0.6247	0.0640	0.7809	0.2174

Table 11.
Analytical method results—Example 4.

Mean and standard deviation for normal variable x_1 remain the same:

$$\mu_{x_1} = 14, \sigma_{x_1} = 0.07$$

Mean and standard deviation for the equivalent normal distribution for the non-normal variables, x_2 and x_3 (**Table 7**):

$$\mu_{x_2} = 3.4147, \sigma_{x_2} = 1.8947, \mu_{x_3} = 0.1997, \sigma_{x_3} = 0.010197$$

$$\alpha_1 = -0.0295, \alpha_2 = 0.9967, \alpha_3 = 0.0751$$

$$\sigma_z = \sqrt{(0.8 * 0.07)^2 + (-1 * 1.8947)^2 + (-14 * 0.010197)^2} = 1.9009$$

$$Z(X) = Z\ (13.9879, 8.2047, 0.20406) = 0.1288$$

$$\beta_{HL} = \frac{Z(X) - \sum_{i=1}^{n} \frac{\partial Z(X)}{\partial x_i} \sigma_{x_i}(u_i)}{\sqrt{\sum_{i=1}^{n} \left(\frac{\partial Z}{\partial x_i} \sigma_{x_i}\right)^2}} = \frac{0.1288 - (-6.4080)}{1.9009} = 3.4388$$

$$P_f = \Phi(-\beta) = 0.00037$$

Table 8 shows that the obtained values for beta and the probability of failure are so close to each other.

The calculated probability of failure exceeds the target probability of failure 10^{-4}.

Example 4

Two systems, system A and system B, each system has three main components and each component with it; the probability distribution type and its parameters are shown in **Table 9**. It is required to determine the probability of failure of each component using FOSM, AFOSM and Monte Carlo simulation [9].

Solution

Formulate a general LSF for each component as:

$$Z(X) = x_1 - x_2$$

Formulate a general LSF for each component see the (**Table 10**).

Analytical Solution

Analytical solution see (**Table 11**).

Author details

Alireda Aljaroudi
St. John's, NL, Canada

*Address all correspondence to: aaa515@mun.ca

References

[1] Aljaroudi A, Khan F, Akinturk A, Haddara M, Thodi P. Risk assessment of offshore crude oil pipeline failure. Journal of Loss Prevention in the Process Industries. 2015;**37**:101-109

[2] Aljaroudi A, Khan F, Akinturk A, Haddara M. Probabilistic performance assessment of fiber-optic leak detection systems. Journal of Offshore Mechanics and Arctic Engineering. 2016;**138**: 021401-1-021401-7

[3] Aljaroudi A, Khan F, Akinturk A, Haddara M, Thodi P, Paulin M. Probabilistic methods for predicting the remaining life of offshore pipelines. Journal of Pressure Vessel Technology. 2017;**139**:041701-1-041701-8

[4] Choi S, Canfield R, Grandhi R. Reliability-Based Structural Design. London, UK: Springer-Verlag; 2007

[5] Haldar A, Mahadevan S. Probability, Reliability and Statistical Methods in Engineering Design. New York, NY, USA: John Wiley and Sons Inc.; 2000

[6] Hines W, Montgomery D, Goldsman D, Borror M. Probability and Statistics in Engineering. New York, NY, USA: John Wiley & Sons; 2003

[7] Hasofer A, Lind N. Exact and invariant second-moment code format. Journal of the Engineering Mechanics Division. 1974;**100**(1):111-121

[8] Kuru M. Reliability based structural and aeroelastic optimization of wing models with high fidelity solvers [Master thesis]. Istanbul, Turkey: Istanbul Technical University Informatics Institute; 2011

[9] Mayda M, Choi S. A reliability-based design framework for early stage of design process. Journal of Brazilian Society of Machinal Sciences and Engineering. 2017;**39**:2105-2120

[10] Nowak A, Collins K. Reliability of Structures. New York: McGraw-Hill; 2000

Chapter 6

NDT Methods for Evaluating FRP-Concrete Bond Performance

Kenneth C. Crawford

Abstract

The long-term bond performance, 15+ years, of FRP-structural systems applied to reinforced-concrete structures is largely unknown and not widely tested. FRP-structural system performance is a function of FRP-concrete bond condition and is subject to deterioration over time. The purpose of this investigation is to test and validate the non-destructive testing impulse-excitation technique to evaluate bond condition of FRP systems applied to concrete structures, in particular concrete highway bridges. The objective is to identify changes in the FRP-concrete bond state by analyzing changes in impulse excitation (impact) frequencies and sinusoid waveforms. Hammer impact tests were performed on two FRP-retrofitted highway bridges in Missouri and a bonded FRP test plate in the laboratory. Signal analysis of recorded impact acoustic emissions was performed on frequencies and waveform damping ratios of bonded and de-bonded FRP material on two bridges and in the lab. The frequencies and sinusoidal waveforms of the bonded and de-bonded FRP material on the bridges had a high degree of correlation to those of the bonded/de-bonded laboratory FRP plate. This investigation confirms the impulse excitation technique to test FRP bond on concrete structures, which provides accurate data on the bonded versus de-bonded FRP-bond condition.

Keywords: bridges, FRP systems, non-destructive testing (NDT), impact-echo, signal analysis, frequencies, damping ratio

1. Introduction

The use of fiber reinforced polymer (FRP) structural systems to strengthen and rehabilitate concrete bridges over the past three decades has given engineers the capability to extend the life and serviceability of highway bridges. When used on reinforced concrete (RC) highway bridges these FRP systems provide unique structural qualities which significantly improve bridge load performance and service life. However, the question is for how long?

For transportation infrastructure owners who have used FRP systems to strengthen highway bridges there is an ongoing concern focused on the long-term durability of these structural systems and how well they will sustain bridge load capacity and performance over time. Bridge owners are faced with two primary questions: (1) how do FRP structural systems on RC highway bridges perform over extended time (15+ years) under the influence of frequent cyclic loading (heavy truck traffic), moisture, and freeze-thaw cycles, and (2) how can accurate field data be obtained on FRP-plate bond conditions to evaluate changes in load performance, to establish cost-effective maintenance procedures, and to determine long-term durability of the applied FRP systems? This concern increases when a highway system has a large number of strengthened bridges in which the load rating of the bridges is dependent on the FRP-concrete bond performance.

With a number of highway bridges having been strengthened with FRP systems over the past several decades in Missouri, Kentucky, Ohio, Canada, Europe, e.g., Slovenia and Macedonia, and the Far East, there has not been an effective field testing system to evaluate and verify the bond condition of large areas FRP material applied to these bridges. While a number of good NDT systems are available to test FRP bonding on concrete structures, these systems are limited in their capabilities to scan extensive areas of FRP material applied to bridge structural members and to produce reliable bond condition data.

The purpose of the investigation in this chapter is to verify field testing capabilities of a light-mobile impact machine for scanning and producing accurate FRP-concrete bond data on bridge structural members retrofitted and strengthened with FRP-structural systems.

The objective of this NDT system, based on the impact-echo principle, is to provide a testing procedure capable of rapidly scanning long lengths FRP plates bonded to bridge structural members, for producing reliable FRP-concrete bond data, and for locating and accurately identifying de-bonded areas, if present [1]. Based on current testing methods, there is a need for an improved cost-effective testing procedure to evaluate FRP-strengthened bridges.

2. Research significance

Quantifying long-term durability of FRP-structural systems applied to concrete structures, and RC bridges in particular, is an on-going issue for the industry and users of such systems. Being able to guarantee FRP systems applied to concrete structures will remain bonded for 20–30–50 years is a challenge. The process to obtain reliable field data proving FRP laminates remain bonded over long periods of time had not been accomplished on a wide scale.

While the impact-echo NDT method has been effective in testing concrete structures for many years its application for evaluating FRP-structural systems applied to RC bridges is not fully developed [2]. The development and testing of the mobile impact machine presented in this paper expands the application of NDT procedures in the field to evaluate bridge CFRP plate-concrete bond conditions. The NDT method presented potentially provides bridge owners a means to rapidly assess bond conditions and to identify associated changes (decreases) in bridge load performance.

2.1 Scope of CFRP applications on RC bridges requiring bond testing

The extent of CFRP material applied to RC highway bridges requiring bond testing is challenging. As an example, in 2001, in one CFRP-bridge strengthening project in the Republic of Northern Macedonia, to meet Eurocode 2-Part 2 (EN 1992-2:2005), 17 slab and girder bridges on the M2 (European Corridor 8, EC-8) were designed and strengthened to NATO MLC 100, having a total of 10,847 m of CFRP plate, including CFRP fiber wrap in bridge structural member shear areas. Two additional slab bridges on the M1 were strengthened in 2002 with 1894 m of CFRP plate.

An example of the length of CFRP material applied to one slab bridge on the M2 is illustrated by bridge B36 at km 67 + 409 in **Figure 1**, showing the distribution of

Figure 1.
Bridge B36 with 792 m in 80 CFRP plates.

the CFRP plates applied in each of two 10 m spans. The center 3.9 m of each 10 m span have 40 S1512 plates parallel to one another in the area of maximum bending moment of the span. The CFRP-concrete bonding in this area is critical for sustaining the designed load capacity for MLC 100 [3].

2.2 FRP-concrete de-bonding mechanisms

The focus of the research presented in this chapter is to develop an effective NDT method to identify FRP-concrete de-bonding on RC structures. The de-bonding of FRP material from concrete is caused by multiple factors. Failure of the adhesive bond of externally applied FRP systems is caused by a number of mechanisms and is referred to as "delamination" or "de-bonding." In their paper, Degala et al., have identified the following parameters as affecting bond interface behavior: concrete tensile strength, adhesive mechanical properties, effective bond length of the FRP, member size and scale, concrete section geometry, structural member loading geometry, degree and rate of loading, FRP retrofit geometry, environmental and mechanical exposure, e.g., moisture and thermal effects, and the mixed mode relationship of de-bonding mechanisms. Any one, or any combination of these mechanisms, can cause de-bonding of FRP material from concrete [4].

The purpose of the NDT method presented in this chapter is to identify FRP-concrete de-bonding in the life cycle of the FRP-strengthened structure and to identify any of the above de-bonding mechanisms potentially affecting RC structure performance to allow early intervention and corrective measures.

3. Experimental program

The scope and purpose of this experimental investigation was to field test and validate a newly developed impulse excitation (impact) machine on CFRP (carbon fiber-reinforced polymer)- strengthened structural members on an RC highway bridge to obtain impact signal data reflecting the condition of the CFRP bond on concrete. The objective was two-fold: (1) to obtain impact data recordings on FRP-strengthened RC bridges and to compare those signal waveforms to the waveforms obtained from the concrete and bonded and de-bonded CFRP plate in the laboratory, and (2) to establish the operability and design limits of the impact machine on bridges. The goal was to verify the capabilities and practicality of such an impact machine in the field on RC bridges, to test FRP laminates on concrete, and to produce reliable FRP-concrete bond data.

3.1 NDT methods for testing FRP-concrete bond

To illustrate the practicality of this impact machine for field testing large areas of CFRP laminates on RC bridges, comparison is made to other NDT procedures used for testing FRP laminates. The use of non-destructive testing (NDT) to evaluate FRP-concrete bond offers a range of effective methods capable of providing data on the behavior and condition of FRP-structural system bonding on concrete structures. There are a number of NDT systems available to evaluate CFRP-structural system bonding on concrete structures, which include acoustic stress waves, infrared thermography, X-ray, microwave, high-frequency radiation, ultrasonics, and laser vibrometry, to detect de-laminations in the FRP-concrete interface regions. Each NDT method offers unique advantages for testing FRP bond under specific conditions [5].

Two systems are considered for comparison to the NDT system presented in this paper. The first system uses pulsed infrared thermography to detect defects in CFRP parts and plates by producing eddy currents generating heat in the part. Heat applied to the test specimen provides spatial temporal distribution of surface temperature whereby defects disturb the heat flow. This system provides high contrast images and is effective in identifying defects in aircraft and automotive parts. It is not economical for scanning large CFRP areas [6].

The second system uses an acoustic-laser technique which generates a sound vibration signature used to identify and characterize defects in CFRP-concrete structural members. Having a good signal-to-noise ratio the system can accurately identify the size of defective areas and has the capability to stand off from the test specimen. It does require set-up of a number of pieces of equipment to generate the acoustic vibrations. The rate in meters per hour at which this system is able to test long lengths of CFRP plates on concrete bridge member is not indicated. Both methods have the capability to identify in detail the precise location and degree of de-bonding of FRP material from concrete [7].

Using acoustic-stress waves from a light-hammer device the NDT method presented in this paper is designed to rapidly scan long lengths of bonded CFRP plates and record changes in impact frequencies. The objective of this system is to quickly provide bonding data over large areas of CFRP plates.

For this investigation the use of the impulse excitation technique with the impact-echo procedure provides an effective and rapid means of data generation to assess the bond condition of large areas of CFRP laminates applied to RC bridges. The advantage of this method over the other three NDT methods is its inherent speed of testing and the volume of bond data produced in the form of digital waveforms.

3.2 Impulse excitation technique

The impulse excitation technique (IET) is a non-destructive testing method for evaluation of elastic and damping properties of materials. This technique is based on the mechanical excitation of a solid body by means of a light impact. For isotropic homogeneous material of simple geometry, such as a prismatic bar, e.g., CFRP plate, the resonant frequency of the free vibration provides information about the elastic properties of the material and its bond to a substrate, i.e., concrete. The amplitude decay of the free vibration is related to the internal friction of the material, i.e., dissipation, or damping, of the vibration energy in the specimen.

Using impulse excitation (impact) on bonded FRP-laminates, the impact frequency, signal amplitude, and the rate of sinusoidal decay (damping ratio) of the harmonic waveform provide information on the condition of the FRP bond to the concrete structural member. Signal analysis of the waveform frequencies, damping ratios, amplitudes, and phase shift provide a detailed and accurate picture of the state of the FRP laminate-concrete bond. The type of harmonic signal produced by the impulse excitation method is a decaying sinusoidal waveform [8].

3.3 Harmonic oscillations

To analyze the waveform data generated by the impulse system, it is useful to consider the characteristics of a decaying harmonic oscillation, shown in **Figure 2**. The natural frequency of a simple continuous harmonic oscillation when the forcing function is zero is expressed by the second order differential in Eq. (1).

$$\mathbf{\frac{dx^2}{dt} + c\frac{dx}{dt} + kx = 0} \tag{1}$$

where c and k are constants related to the impulse system and material tested, and $\mathbf{c \geq 0, k \geq 0}$. When the harmonic oscillation is driven by forcing function $f(t)$ on mass m then $f(t)$ is expressed in Eq. (2).

$$\mathrm{m}\ \frac{\mathrm{dx}^2}{\mathrm{dt}} + \mathrm{c}\frac{\mathrm{dx}}{\mathrm{dt}} + \mathrm{kx} = \mathrm{f(t)} \tag{2}$$

The general solution of $f(t)$ for a dampened harmonic waveform is given by Eq. (3).

$$f(t) = \mathrm{A}\, e^{-\zeta \omega n t}\ (\cos{(\omega_{\mathrm{d}} - \phi)}) \tag{3}$$

where A is initial amplitude, ω_n is the natural frequency in rad/s, ϕ is phase angle, ζ is the damping ratio on ω_n. The dampened frequency ω_d is the natural frequency ω_n modified by the dampening ratio.

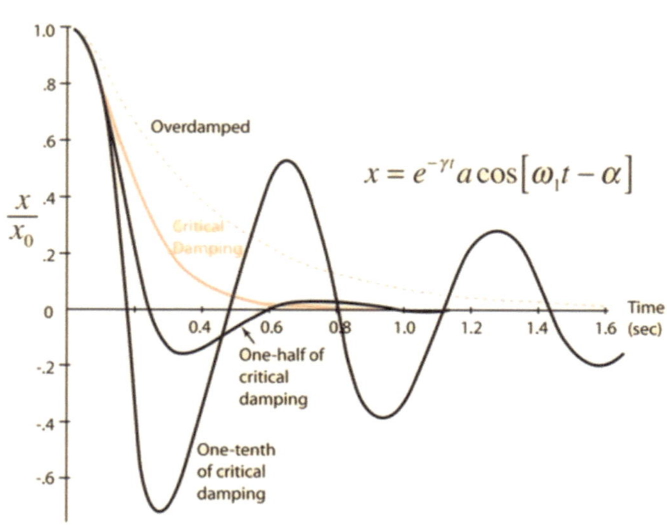

Figure 2.
Exponentially decaying harmonic oscillation @ varying rates.

3.4 Damping ratio

The rate of exponential decay in a harmonic oscillation is expressed by the damping ratio, a system parameter denoted by zeta (ζ) that characterizes the frequency response of the harmonic oscillation. The simplest example is a mass m suspended from a spring with a spring constant k, with a damping ratio c. When a single forcing function $f(t)$ is applied to the mass $x(t)$ the result is a harmonic oscillation with a unique rate of decay determined by the damping ratio, which is a function of the internal resistance of the system driven by the forcing function.

The damping ratio expressed by zeta (ζ) is the ratio of actual damping to critical damping cc, Eq. (4), such that ζ = actual damping/critical damping cc, where $cc = 1$. The damped circular frequency ω_d is the natural frequency ω_n reduced by the factor $\sqrt{1-\zeta^2}$.

$$\omega_d = \sqrt{1-\zeta^2}\omega_n \qquad (4)$$

3.5 Fourier transform of impact signal

The impact oscillation frequency from the forcing function $f(t)$, where $f(t) = Ae^{-\zeta\omega nt}$ $(\cos(\omega_d - \phi))$ is represented by the Fourier transform of the damped cosine in Eq. (5).

$$f(\omega) = (j\omega + a)/(-\omega^2 + 2j\omega a + \omega_o{}^2 + a^2) \qquad (5)$$

where $a = \zeta\omega_n$ and $\omega_o = \omega_n$. The Fourier transform provides the frequencies of the time domain signals $f(t)$. As the NDT device moves across the surface of the CFRP plate changes in impact signal frequencies can be identified. A change in the signal frequency indicates a change in the CFRP-concrete bond condition. If the impact frequencies across the plate remain relatively constant, the plate is fully bonded.

4. Impact machine

The latest impact machine, **Figure 3**, is a 5 × 17 cm single 3D-printed frame made with polylactic acid (PLA) plastic by fused filament fabrication, is used for its strength and workability. PLA, or polylactide, is a biodegradable and bioactive thermoplastic aliphatic polyester derived from renewable resources and provides a

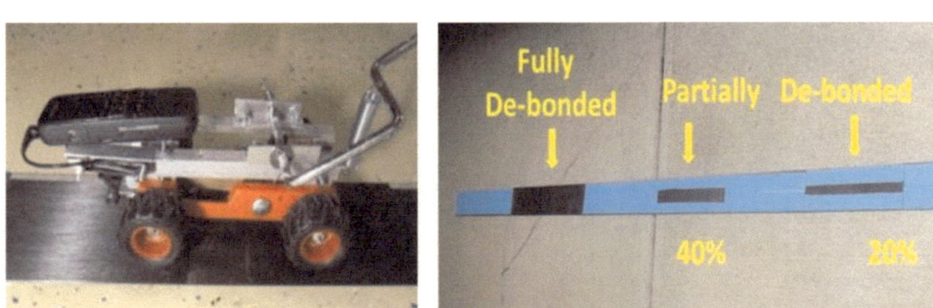

(a) (b)

Figure 3.
(a) Mobile impact machine and (b) laboratory CFRP test plate.

strong structural frame which attenuates vibrations and sound, ideal for this impact machine application. The machine has a single 10 cm lever (cherry wood) with one 10-24 screw for the impact hammer, driven by a four-pin actuator wheel. The machine produces 25 impacts per meter, with a 0.4 pound impact force. The impact sound wave signals generated are recorded with a microphone and digital data recorder mounted on the machine.

5. Experimental procedure

The experimental investigation was conducted in two phases. Phase one involved testing in the laboratory with the impact machine, **Figure 3a**, on a bonded, de-bonded, and partially de-bonded CFRP plate, **Figure 3b**, and on the concrete to which the CFRP plate was bonded. The purpose of obtaining impact waveforms on the laboratory concrete was to compare them to the waveforms on the bridge concrete. Similarly the impact waveforms obtained from the laboratory bonded and de-bonded CFRP plate are used to compare to the bonded and de-bonded CFRP material on the bridges. A partially de-bonded CFRP plate was simulated in the laboratory with 40 and 20% de-bonding of the test plate, **Figure 2b**. Multiple passes of the impact machine in the laboratory on the concrete, bonded, de-bonded, and partially de-bonded CFRP plate were made with the data recorded on a Sony M2 digital recorder. The recordings were displayed in a Tektronix 475 oscilloscope and video recorded. The video recordings were broken down into still pictures of the waveforms, from which the impact signal frequencies and damping ratios were measured. The frequencies were obtained from the inverse of the time displayed on the oscilloscope screen. The damping ratios were measured with curve fitting of the decayed sinusoid waveforms.

Phase two used a similar impact test procedure in the field on the Creasy Springs Lane bridge and the Coats Lane bridge. Impact tests were run on the bridge concrete first, and then multiple impact runs were performed on various beams of the RC channels. While not all beams were tested, randomly selected beams were tested, generally those where the paint was gone exposing the CFRP material. Performing impact tests on painted CFRP material will require a modified (stiffer) impact machine to produce stronger signals to reflect the CFRP-concrete bond conditions.

The laboratory test consisted of running the impact machine, in a number of cycles, first on concrete and then on a bonded, de-bonded, and partially de-bonded CFRP plate to produce impact signal waveforms representing the three CFRP-concrete bond conditions. To obtain signal waveforms on RC bridges, field tests were performed on two CFRP-strengthened county highway bridges in Missouri. The test involved running the impact machine on the bottom side of beams on the bridge channel sections. The two bridges tested were the 39 foot (12 m) Coats Lane bridge and the 19 foot (6 m) Creasy Springs Road bridge, both on county roads in Boone County, near Columbia, Missouri. Both bridges were constructed in the early 1970s with precast reinforced concrete channel sections with a four in (10 cm) deck slab running the length of the bridge. The channel beams were strengthened with CFRP plate and fabric material to increase the bridge load rating from 15 to 20 tons. The bridge channel sections (**Figure 3**) have a 10 cm CFRP plate on the bottom of each beam and 60 cm fabric wrap with 12 wraps per beam. The bridges were strengthened in the early 2000s [9].

The Creasy Springs Road bridge, **Figure 4**, constructed with precast RC channels like Coats Lane, has the same CFRP plate and fabric configuration on the channel beams as the Coats Lane bridge.

On the Creasy Springs Road bridge impact testing, **Figure 5**, was done on the second beam in the first channel (west side) and then on the third, fourth and sixth beams as far as the extension pole could reach over the water. It is noted CFRP de-bonding was found on the second beam, **Figure 5c**.

The Coats Lane bridge, **Figures 6** and 7, has eight RC channels, two beams per channel. Testing was performed on the first channel second beam (north west side), **Figure 6a**, and then on beams six, seven, and eight in areas were the paint had flaked off, **Figure 6c**.

5.1 Experimental results

Impact waveform data was compiled on results of testing the CFRP plate in the laboratory and on the Creasy Springs Road and Coats Lane bridges in Missouri. Selected samples of the hundreds of impact data points (signal waveforms) are shown in **Figures 8–13**. The waveforms obtained in the laboratory for concrete, bonded, and de-bonded CFRP material are compared to the waveforms obtained on the concrete and CFRP material on the two bridges. Comparisons are made of the frequencies and damping ratio in the three scenarios.

Figure 4.
Creasy Springs Road bridge.

(a) (b) (c)

Figure 5.
Creasy Springs Road bridge—impact testing on channel beams. (a) Second and third beams, (b) testing fourth beam, and (c) CFRP de-bonding second beam.

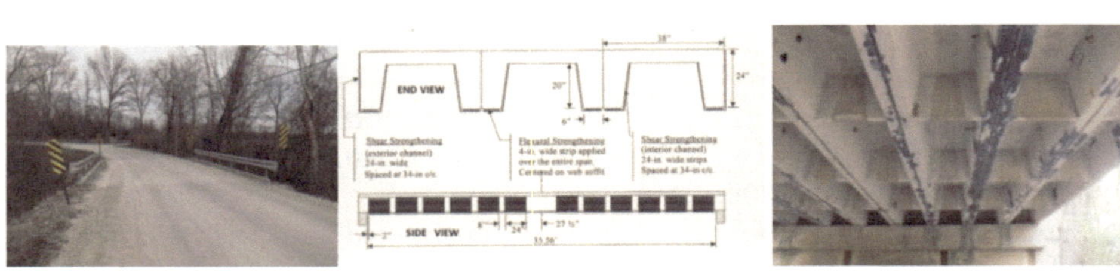

Figure 6.
Coats Lane bridge (Boone County, MO) with CFRP configuration on channel beams.

DOI: http://dx.doi.org/10.5772/intechopen.88105

(a) (b) (c)

Figure 7.
Coats Lane bridge—testing channel beams. (a) Second beam, first channel, (b) impact testing, (c) beams 6–8.

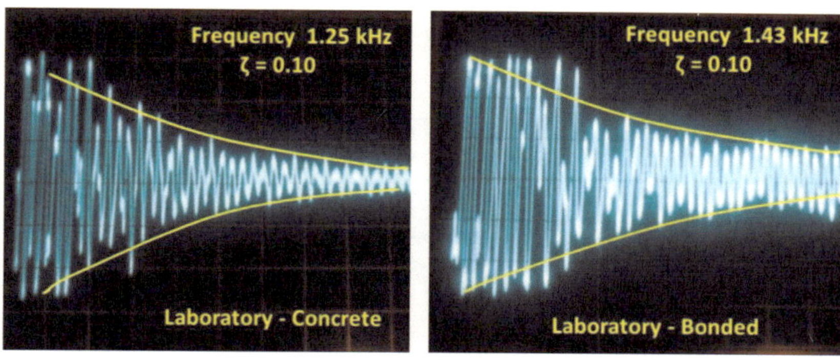

Figure 8.
Impact waveforms—laboratory—concrete and bonded CFRP plate.

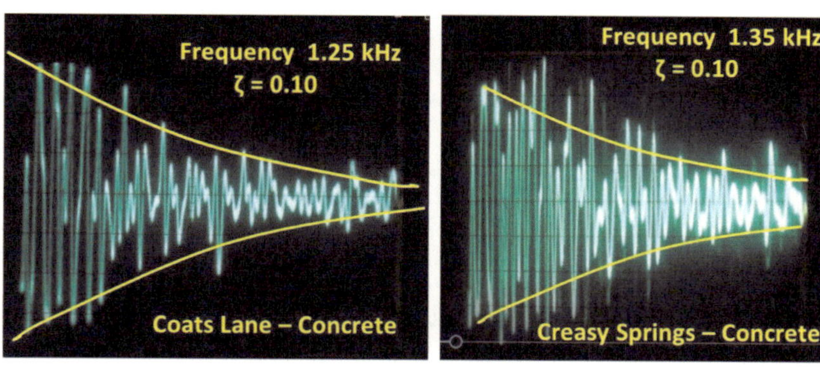

Figure 9.
Waveforms—concrete—Coats Lane and Creasy Springs bridges.

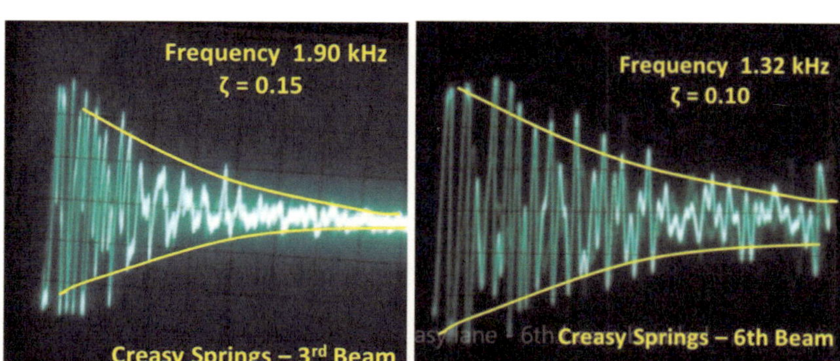

Figure 10.
Waveforms—Creasy Springs bridge—bonded CFRP plate.

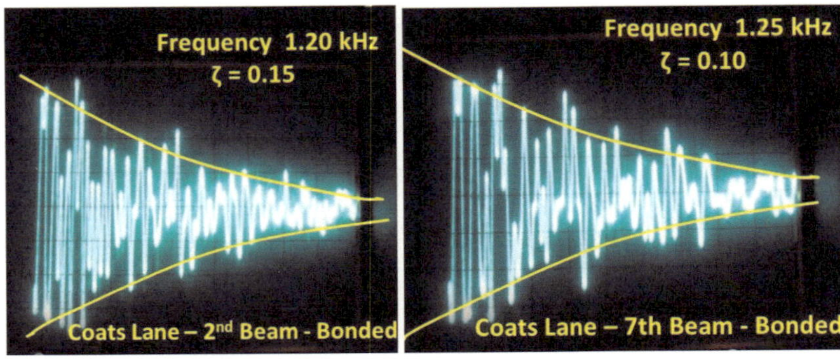

Figure 11.
Waveforms—bonded CFRP plates on Coats Lane channel beams.

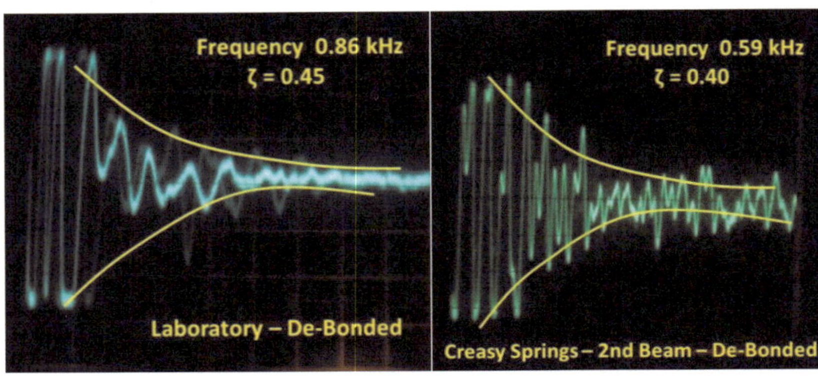

Figure 12.
Waveforms—de-bonded CFRP plate—lab and Creasy Springs bridge—second beam.

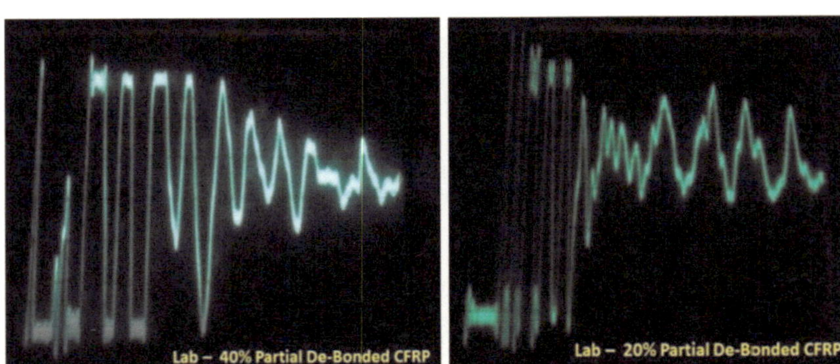

Figure 13.
Waveforms for partially de-bonded CFRP plate—40 and 20%.

The results of the impact tests on the two bridge and the laboratory are summarized in Appendix A, **Table 1**.

The laboratory concrete waveform produced a frequency of 1.25 kHz and a damping ration of 0.1. The concrete waveform on the two bridges also produced a frequency of 1.25 kHz, plus or minus 15 hz, and a damping ratio close to 0.1. The CFRP bonded frequency in the laboratory was 1.43 kHz, with a damping ration of 0.1. The CFRP bonding frequency obtained on the two bridges was 1.90 kHz on

Laboratory CFRP test plate					
Acoustic signal	Concrete	Plate bonded		Plate de-bonded	
Frequency (kHz)	1.25	1.43		0.86	
Damping ratio ζ	0.10	0.10		0.45	
Creasy Springs bridge					
Acoustic signal	Concrete	Second beam		Third beam bonded	Sixth beam bonded
		Bonded	De-bonded		
Frequency (kHz)	1.35	—	0.590	1.90	1.32
Damping ratio ζ	0.10	—	0.40	0.15	0.10
Coats Lane bridge					
Acoustic signal	Concrete	Second beam bonded	Sixth beam bonded	Seventh beam bonded	Eighth beam bonded
Frequency (kHz)	1.25	1.20	1.32	1.25	1.28
Damping ratio ζ	0.10	0.15	0.10	0.10	0.15

Table 1.
Acoustic impact data: laboratory test plate and RC highway bridges.

Creasy Springs and 1.32 kHz on Coats Lane, each with a damping ratio of 0.1. Impact test on Coats Lane channel beams two, six, and seven had frequencies of 1.20, 1,25, and 1,25 kHz, plus or minus 15 hz, with all damping ratios close to 0.1. The frequency and damping ratio of the CFRP de-bonding in the laboratory and the CFRP de-bonding found on beam two on the Creasy Springs bridge, **Figures 5c** and **9**, were closely related, with 0.895 and 0.590 kHz respectively, and damping ratios of 0.3 and 0.4. Partial de-bonding on the laboratory test plate was tested with the resulting waveforms shown in **Figure 13**, for 40 and 20% de-bonding. The impact machine can detect partial de-bonding. It is noted the relationship between the frequency and damping ratio changes as the de-bonded area of the CFRP plate become smaller, but still having a distinctive waveform produced which indicates de-bonding, **Figure 11**. The impact machine cannot indicate the degree of partial de-bonding. Other NDT methods, such as thermal imaging, are required to determine the degree and extent of partial CFRP de-bonding.

The application of the mobile impact machine, **Figure 3a**, with its data recorder proved successful in detecting bonded and de-bonded CFRP material in the laboratory and in the field on RC highway bridges. The waveforms for bonded and de-bonded CFRP material produced in the laboratory were consistent and similar to the bonded and de-bonded waveforms produced on the two RC bridges. This indicates the concept of using the impact-echo NDT method with the impact machine to test FRP-bonding on RC bridges and producing bond condition data is valid. It was noted the impact data obtained in the field on the two bridges had a relatively high degree of noise in the waveforms. Improvements in the impact system and data recording method, with noise filters, can improve the impact signal to noise ratio, producing a relatively cleaner and more distinguishable and measurable waveform, denoting CFRP-concrete bond condition.

6. Conclusion

The main purpose of this investigation was to test and validate the application of the NDT impulse-excitation technique to measure bond condition of

FRP-structural systems applied to concrete structures. This impact method produces acoustic emissions from bonded FRP material displaying a decaying sinusoidal waveform with unique frequencies and damping ratios. The frequencies and damping ratios identify two FRP bond state conditions: bonded and de-bonded. The de-bonded state has lower frequencies with higher (faster decay) damping ratios. The change in impact frequencies and the rate of decay in the damping ratio are the two primary parameters identifying changes in the FRP-concrete bond state.

Two FRP-retrofitted highway bridges in Missouri were tested in this investigation. Tests were made on a bonded/de-bonded FRP plate in the laboratory to provide control and a reference for the bonded and de-bonded states. Impact data results from the two bridges matched the laboratory test data for the bonded and de-bonded conditions. Impact tests on the bonded FRP material on the bridges had frequencies in the 1.25 kHz range with damping ratios of 0.10. These values matched the 1.20 kHz frequency and 0.10 damping ratio for the bonded FRP material in the lab. The de-bonded area on one bridge had a frequency of 0.85 kHz and damping ratio of 0.40 matching the de-bonded FRP values in the lab.

The NDT impulse excitation technique testing the bond of FRP-structural systems applied to concrete structures produces clear decaying sinusoidal waveforms. Signal analysis of the waveform frequencies and damping ratios confirmed a direct correlation to the FRP-concrete bond condition. Field and laboratory test results in this investigation validated the impulse excitation technique applied to bonded FRP material accurately measures the bond state between the FRP material and the concrete structure. The use of the impulse excitation technique in this investigation to evaluate FRP-bond condition has broader applications for the non-destructive testing of FRP-retrofitted concrete structures and related fields in civil engineering.

Acknowledgements

The author thanks Wabash College for their support and the work performed by the Wabash College 3D Printing and Fabrication Center, supported through a Ball Brothers Foundation Venture Fund Grant, and for the facilities and instrumentation support for their 3D-printing of the impact machine frame used in this investigation.

Conflict of interest

The author declares no conflict of interest.

Appendix A

See Appendix A, Table 1.

Author details

Kenneth C. Crawford[1,2,3,4]

1 Crawford Technologies and Applications, LLC, Crawfordsville, Indiana, USA

2 Institute for Bridge Reinforcement and Rehabilitation, Bloomington, Indiana, USA

3 American Concrete Institute (ACI), USA

4 International Institute for FRP in Construction (IIFC), Canada

*Address all correspondence to: ken.crawford@ibrr.org

References

[1] Crawford K. NDT evaluation of long-term bond durability on CFRP-structural systems applied to RC highway bridges. International Journal of Advanced Structural Engineering. 2016;**8**(2):161-168

[2] Sansalone M, Street W. The impact-echo method. NDTnet. 1998;**3**(2):3-5

[3] Crawford K, Nikolovski T. The application of ACI 440 for FRP system design to strengthen 19 concrete highway bridges on European corridor 8. In: 8th International Conference on Fiber Reinforced Polymer for Reinforcement of Concrete Structures; Paper 15-4; 2007. pp. 570-571

[4] Degala S, Ramanathan K, Rizzo P, Hatties K. Structural health monitoring of CFRP reinforced concrete one way slabs by acoustic emission. In: Advanced Composite Materials in Bridges and Structures, Proceedings ACMBS-V; 2008

[5] Banthia N, Abdolrahimzadeh A, Demers M, Mufti AA, Sheikh SA. Durability of FRP-concrete bond in FRP-strengthened bridges. ACI Concrete International. 2010;**32**(8): 45-51

[6] Maier A et al. Non-destructive thermography analysis of impact damage on large-scale CFRP automotive parts. Materials. 2014;**7**:413-429

[7] Büyüköztürk O. Remote detection of de-bonding in FRP-strengthened concrete structures using acoustic-laser techniques. In: Nondestructive Testing of Materials and Structures, Proceedings NDTMS; Part 1; 2011. pp. 19-24

[8] Crawford K, Chapman C, Porter L. Low frequency analysis of CFRP-laminate bond on RC bridges. In: Proceedings of 7th International Conference on Advanced Composite Materials in Bridges and Structures (ACMBS-VII); August 2017; Vancouver, BC, Canada; Paper 065; 2017

[9] Alkhrdaji T, Thomas T, Nanni A, Huck M. Strengthening of off-system bridges using FRP composites. OAI: CiteSeerX. Psu: 10.1.1506.4417; 2014

Chapter 7

Temperature Diffusivity Measurement and Nondestructive Testing Requiring No Extensive Sample Preparation and Using Stepwise Point Heating and IR Thermography

Dmitry Yu. Golovin, Alexander G. Divin, Alexander A. Samodurov, Alexander I. Tyurin and Yuri I. Golovin

Abstract

This chapter describes a modification to the laser flash method that allows determining temperature diffusivity and nondestructive testing of materials and constructions without cutting samples of predefined geometry. Stepwise local heating of the studied object surface at a small spot around 0.1 mm radius with simultaneous high temporary-spatial resolution infrared (IR) filming of the transient temperature distribution evolution with a thermal camera provides a wide range of possibilities for material characterization and sample testing. In case of isotropic and macroscopic homogeneous materials, the resulting transient temperature distribution is radially symmetric that renders possible to improve temperature measurement accuracy by averaging many pixels of the IR images located at the same distance from the heating spot center. The temperature diffusivity measurement can be conducted either on thin plates or on massive samples. The developed emissivity independent in plain IR thermographic method and mathematical algorithms enable thermal diffusivity measurement for both cases with accuracy around a few per cent for a wide range of materials starting from refractory ceramics to well-conducting metals. To detect defects, the differential algorithm was used. Subtracting averaged radial symmetric temperature distribution from the original one for each frame makes local inhomogeneities in the sample under study clearly discernible. When applied to crack detection in plates, the technique demonstrates good sensitivity to part-through cracks located both at the visible and invisible sides of the studied object.

Keywords: laser stepwise point heating, transient IR thermography, temperature diffusivity measurement, nondestructive testing, crack detection

1. Introduction

Active and passive methods of thermal testing and evaluation have arisen and developed for more than a century [1, 2]. Great progress in thermophysical research has been achieved in the past few decades due to the development of laser and noncontact infrared (IR) thermography technologies [3–5]. The active mode of thermography is most often implemented using optical excitation (stimulation) generated by laser or high-powered halogen lamps with subsequent or simultaneous recording of transient temperature distribution with a high spatial-temporal resolution thermal camera [6–9]. Modern thermographic methods employing mathematical models and thermogram analysis techniques developed in recent years allow extracting a lot of useful information about the state of various objects—from small components of microelectronics [10, 11] to large products [12–15], artworks [16, 17], and engineering structures [14, 18, 19]. They can also be used to conduct nondestructive testing of materials [20–27], determine their thermal physical characteristics (TPC) [28–31], and determine the degree of corrosive damage [32]. Thermography is also used successfully for monitoring natural objects in the environment [33], biomedical research and noninvasive diagnosis of various human diseases [9, 34, 35], and in sports science [36].

For active nondestructive testing (NDT) and TPC measuring, two measurement methodologies are used: steady and transient [3–9, 23, 27–30, 37]. The old steady-state methods (in particular, guarded hot plate (GHP) and heat-flow meter (HFM) [28]) are still used in practice, though they have several disadvantages because the distribution of heat flux should be sufficiently close to the planned asymptotics without significant uncontrolled heat leakage from the measuring cell. These methods require preparation of samples of specific geometry and a long time to achieve steady state.

It greatly complicates the observance of the boundary conditions on the free surfaces of the sample under study adopted in the computation scheme and increases the effect of various drifts on the measurement accuracy. Installations of this type require special measures of thermal insulation and thermal stabilization of the sample, heater, refrigerator, and measuring chain, providing low thermal resistance between the heater, sample, refrigerator, and temperature sensors. Due to the above, installations for the stationary methods are cumbersome, require fairly large samples of strictly defined geometry and thermal conductivity can be measured only within a narrow range.

Transient methods are more diverse. They use various temporal profiles of incoming heat flow such as Dirac delta function, step or periodic ones [18, 23, 27, 28, 38]. The latter mode is used by thermal wave interferometry (TWI) technique, which opens additional possibilities related to synchronous signal detection referred to as lock-in technique like accuracy improvement, reduction of drift impact, and gathering information from two independent channels, namely signal amplitude and phase [38–40]. However, due to the necessity for oscillations to be close to asymptotic, this method exhibits some of the abovementioned drawbacks of stationary methods. Another disadvantage of this technique is the high cost of the required hardware.

Among nonharmonic transient techniques of thermal stimulation, laser flash technique (LFT) proposed back in 1961 by Parker et al. [41] is the most widely used up to date. It can be easily adopted for testing plate samples of homogeneous [42], anisotropic [43, 44], porous [45], translucent [46] solids as well as liquids [47–52]. LFT is supported by commercially available equipment [53, 54] operating according to established standards [55–57].

Further advances in LFT result in variants employing thermal wave propagation not only in normal direction as in classic LFT [41], but also in plane techniques [44, 58–62]. It uses point heating to implement the radial diverging thermal wave technique for immobile objects [44, 58–61] and for moving ones [62]. Heating from ring-shaped area in the radial converging thermal wave technique provides better signal/noise ratio and reduces overheating in the irradiated area [63–67]. These variations of LFT allow TPC evaluation and NDT in case of one-side access to the object but use short pulse heating.

The conventional laser flash technique and appropriate equipment has a number of undeniable advantages including high accuracy and good reproducibility of measurements, short testing times (~1 s), optional control over surrounding gas composition, and temperature in wide range (from −120 to 2800°C). However, it has several immanent drawbacks. LFT testing requires specifically prepared samples of the certain geometry. Usually, it is a disk with a diameter of 12–25 mm and thickness ~1 mm. The experiments are costly and time-consuming, and they require two-side access to the sample. This prevents using the method on massive or unprepared samples as well as on finished goods in operation or the ones that can be accessed from one side only.

So, the purpose of the work is development of all-round express technique for temperature diffusivity (TD) *a* measurement in various materials and objects requiring no extensive sample preparation as well as technique for detecting various inhomogeneities and defects in the material. The original algorithms and software were designed and implemented.

2. Methods and materials

The proposed method [68] is based upon thermal imaging of transient temperature field induced by local spot, further referred to as point heating. Original algorithms and software are used for processing IR images recorded by a thermal camera.

Thermal cameras are used in various applications [2, 3, 21, 23, 24] and have many advantages for recording temperature distribution evolution in a studied object. Cameras have a number of advantages over local temperature sensors such as thermocouple elements or thermistors, beside their evident ability to register temperature in a great number of points remotely and simultaneously. High accuracy of data spatial localization, no problems related to thermal inertia, additional thermal capacity, heat sink, and boundary thermal resistance of the gauge should be mentioned as the most prominent ones.

Meanwhile, it should be noted that the lack of accurate information concerning optical absorption and radiation coefficients of the studied object and their variations along the surface render precise measurement of absolute temperatures and heat fluxes in an object difficult in case of laser heating and IR camera recording. But, similar to LFT, the proposed technique requires no absolute values of temperature because the nature of temperature diffusivity a is determined by characteristic length r and time t of thermal wave propagation from hot spot as could be seen from its physical meaning, m^2/s units and dimensionless Fourier similarity criterion $F_0 = at/r^2$. The only important reservation concerns the uniformity of the coefficient between the absolute temperature and acquired signal all over the studied surface. By the way, it is also true for classic LFT that determines the time of reaching the half of temperature maximal value at the back side of the sample of known thickness but not absolute value of the temperature itself.

The proposed approach utilizes step-like temporal profile of point local heating of the surface by laser beam with Gauss radius r_0 = 0.1–0.3 mm and simultaneous filming of transient temperature field induced by this heating at the object surface by an IR camera. In our experiments, laser heating has constant power during all testing time (step-heating thermography) like [69–72] whereas classic LFT uses short pulse heating [44, 58–66]. It significantly reduced the temperature in the hot spot, while increasing it in the surrounding inspected area and lowering requirements to laser output power. The following computer processing of radially symmetric thermal images using the proposed models and algorithms allows determining the value of a in homogeneous materials and detecting and characterizing inhomogeneities and defects owing to local divergence of temperature distribution from axial symmetry. Two variants of the method are considered in the paper namely one for bulk and the other for thin plate objects. They are related to standard LFT but do not require extensive sample preparation.

Verification of the described approach to measurement of a has been done for a number of materials (see **Table 1**) with significantly different thermal and optical properties. A 2-mm-thick square steel plate of 100 × 100 mm was used as a sample. Its TD was determined using a 2D model of thermal wave. All other samples were massive disks with diameter > 30 mm and height > 25 mm. They were studied using a 3D model.

A solid-state Nd:YAG laser unit LSR445CP-FC-10 W with diode excitation, maximal output power of 12 W, 445 nm wavelength, and controllable intensity was used as a heater. Digital IR filming of the object surface was done using an FLIR A35sc thermal camera having 320 × 256 pixels matrix of VO_x bolometers with maximal sensitivity in 7.5–13 mkm range, angular resolution 2.78 mrad, sensitivity ~0.05°C in the range from −20°C to +550°C and frame rate up to 60 Hz. Individual frames like the ones shown in **Figure 1** were converted to two-dimensional temperature maps. To suppress signal inhomogeneities originating from differences in

Material	Composition	Short name
Dense stabilized ceramics ZrO_2:MgO	ZrO_2 (98%). MgO (2%).	Ceramics
Composite based on phenol-formaldehyde polymer	Phenol-formaldehyde polymer (50%) + mica (15%), carbon black (15%), kaolin (10%), talc (5%), and cellulose (5%).	Composite
Cured rubber	Synthetic rubber (60%), carbon black (30%), ZnO (8.5%), and S (1.5%).	Rubber
Teflon (white)	Tetrafluoroethylene	Teflon-w
Teflon (black)	Tetrafluoroethylene + carbon black (40%)	Teflon-b
Gypsum	$CaSO_4 \cdot 2H_2O$ (at least 90%); SiO_2, Al_2O_3, Fe_2O_3, and other (no more than 10%).	Gypsum
Mild carbon steel "Steel 40X" (Analogue of mild carbon steel AISI 5140)	C (0.36–0.44%), Si (0.17–0.37%), Mn (0.5–0.8%), Ni (≤0.3%), Cr (0.8–1.1%), the rest is Fe	Steel

All per cents are mass ones.

Table 1.
Tested materials.

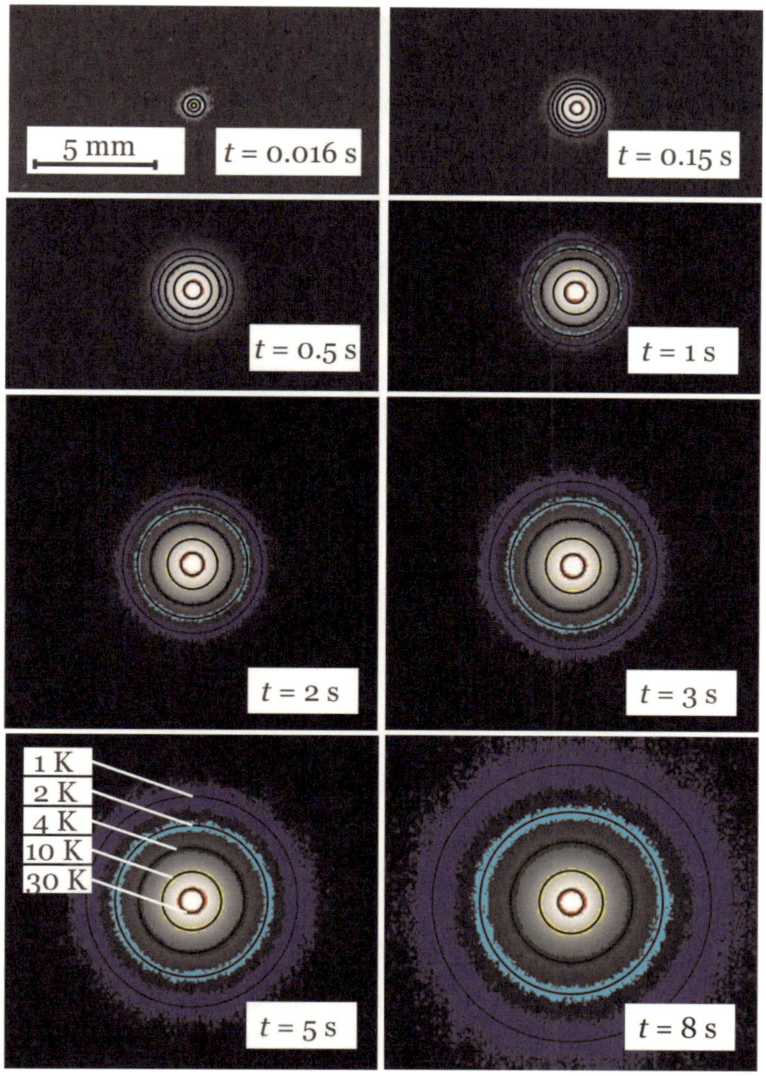

Figure 1.
Certain frames of IR film with added isotherms showing thermal wave propagation resulting from local heating of a ceramic sample. Colored rings correspond to 1 ± 0.25 K, 2 ± 0.25 K, 4 ± 0.5 K, 10 ± 1 K, and 30 ± 5 K.

local surface emissivity, the initial temperature T_0 before the heating onset has been subtracted from temperature T_h recorded during heating for every pixel. From now on, the temperature excess $T = T_h - T_0$ is the signal to be analyzed. The next step is determining position of effective center of the heating and averaging the values of T over the angle φ (see **Figure 2**) for each given distance r from the center to reduce noise, impacts of inhomogeneities, etc. Averaging T over φ is executed in ring-shaped areas limited by radii $r - \Delta r$ and $r + \Delta r$, where Δr is defined by image pixels size and is about 2–10% of r. Usually, $\varphi = 360°$ in TD measurements, and in NTD testing, $\varphi < 360°$ to exclude the area with suspected defects.

The detailed step-by-step raw data processing algorithm is as follows:

1. converting native data format of recorded film in a series of two-dimensional arrays of temperatures;

Figure 2.
Raw data processing scheme. Averaging the temperatures from a set of pixels located at the same distance r in a fragment of last frame from ***Figure 1****.*

2. determining the time of heating onset (in case of the lack of hardware synchronization);

3. averaging each pixel over time before this moment and forming the base frame;

4. pixel-by-pixel subtraction of this base frame from each following frame to suppress the impact of local inhomogeneities of studied surface optical properties;

5. determining the effective center of the heating;

6. averaging temperatures of all the pixels located at the same distance from the heating center. In case of NDT, one needs to search and exclude the areas significantly differing from the mean value to avoid artifacts. The temperatures averaged in the above procedure form the target radial temperature distribution $T(r)$ for the frame. On processing all the frames, one gets $T(r, t)$ distribution; and

7. approximating experimental radial temperature distribution $T(r)$ by function (2) using the Nelder-Mead method [15] in case of 3D model or using linear regression in appropriate coordinates for 2D model. Spatial and temporal limits of the area for data to be approximated are determined by the assumptions made in building model and approximating function. From below, the data related to $t < 3t^*$ and $r < 2r_0$ are ignored, and from above, the areas are restricted by a stronger limit of (a) T should be above 1/10 of maximal accepted value and (b) T should exceed tripled root-mean-square value of the noise.

Distributions $T(r, t)$ obtained for two macroscopically homogeneous defect-free samples using 2D and 3D techniques as described above are presented in **Figure 3**. It should be noted that for thin plates tested in 2D mode, IR camera can be located either at the same side or at the opposite side relative to the incident laser beam (see inset in **Figure 3**). It does not matter principally, but slightly affects details of implementation.

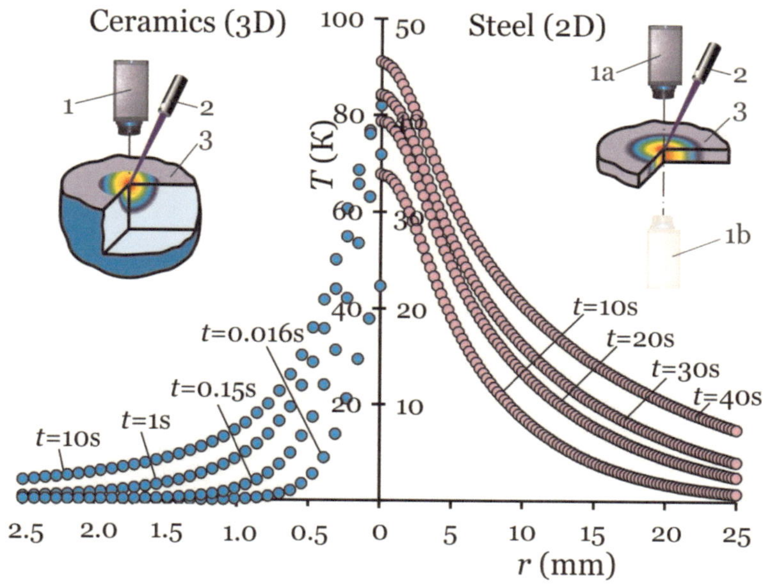

Figure 3.
Experimental radial dependencies of temperature excess T in massive ceramic (3D) and steel plate with a thickness of δ = 2 mm (2D) samples at different times t since the heating onset. r is the distance to the heating center. 1—IR camera; 1a u 1b—two variants of camera positions; 2—laser; and 3—sample. Insets show the experiment layout and color-coded temperature distributions near the heating center.

3. Thermal diffusivity measuring: 3D model

Point heating at the surface of a massive isotropic and homogeneous sample induces radially symmetric spherical thermal wave in the semi-infinite space. The measurement duration τ should be at least an order of magnitude higher that characteristic time $\tau_F = (r_0)^2/a$ for a given material and Gaussian radius of heating spot r_0, so that the thermal wave has sufficient time to propagate to the distance $r > (3–5)r_0$. For $r_0 \sim 0.1$ mm, minimal value of τ is usually below a second for all materials, and for $r_0 \sim 0.3$ mm, it is tenths of a second for metals, a few seconds for ceramics and the like, and up to a few tens of seconds for materials with the lowest values of TD coefficient like highly porous medias and some polymers. Provided condition $\tau \gg \tau_F$ is met, particularly radial distribution of energy in incident beam becomes insignificant at the distances from the heating center $r > (2–3)\ r_0$, and temperature distribution in homogeneous semispace can be considered spherically symmetric up to the times limited only by thermal wave approaching the boundaries of the heated object. For the above values of temperatures, times and distances, the heat exchange at the free surface can definitely be neglected.

Even for isotropic homogeneous materials without temperature dependence of TPC in adiabatic approximation, that is neglecting heat exchange at the surface, and for energy of incident laser beam transferred to the sample not just at the surface but in a sphere with radius R, so that the heat flux through it is uniform and Q in total, precise solution of Fourier heat transfer equation is quite complex and hardly applicable for experimental data analysis [73]:

$$T(r,t) = (Q/2\pi\lambda r)\left[\mathrm{erfc}\left((r-R)/2(at)^{1/2}\right) - \exp\left[(r-R)/R - at/R^2\right] \mathrm{erfc}\left[(r-R)/2(at)^{1/2} + (at)^{1/2}/R\right]\right] \quad (1)$$

where $\mathrm{erfc}(x) = \frac{2}{\sqrt{\pi}}\int_x^\infty e^{-y^2}dy$ is the complementary error function and λ is the material thermal conductivity. Besides, any accurate solution as in Eq. (1) depends upon poorly known energy distribution in the heating spot and optical properties of material such as absorption and stray light factors. However, for $r \gg r_0$ and $t \gg t^* = r_0^2/a$, the temperature distribution $T(r, t)$ can be approximated with accuracy around 1% by function.

$$T(r,t) = Br^{-1}\mathrm{erfc}\left[0.5(r - r_0)(at)^{-1/2}\right] \quad (2)$$

where B is some constant depending upon supplied energy, r_0 and λ.

As could be seen from Eq. (2), temperature distribution approaches asymptotic $1/r$ in time t, determined by the value of a. To obtain the value of a from experimental data manually, it is rational to redraw distribution $T(r, t)$ in (T_r, $(r - r_0)t^{-1/2}$) coordinates and then approximate if with a function $y = D{\cdot}\mathrm{erfc}(A{\cdot}x)$, varying only scales of the axes. As follows from Eq. (2), the value of $A = 0.5a^{-1/2}$ found in this way allows determining $a = (4A^2)^{-1}$. In computer-aided analysis, experimental distributions have been fitted using Nelder-Mead method minimizing sum of squared discrepancies either directly using the function Eq. (2) or in the coordinates described above. Both variants give the results of similar quality. In real experiments, accuracy of T recording deteriorates with growing r due to finite temperature resolution of the camera and growing role of noises of various nature, so that the reasonable testing duration lies in the range $3\tau_F \lesssim \tau \lesssim 10\tau_F$. Raw experimental data processed as described above is shown in **Figure 4**.

Values of a calculated as described above for each of eight individual tests, standard deviation δa of the sampling within each test, averaged value a_8, and standard value a^* for each of the tested materials are presented in **Table 2**.

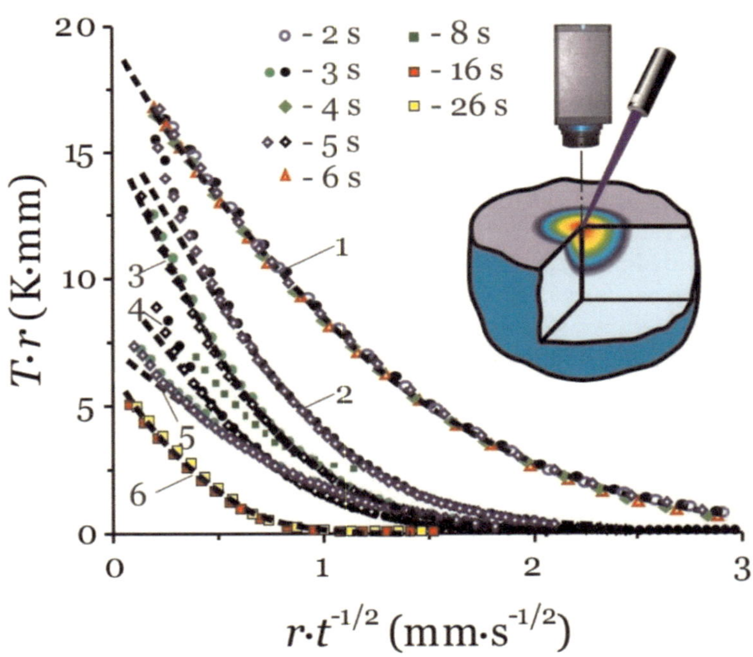

Figure 4.
Experimental temperature dependencies on radius and time in coordinates. $T_r = f[(r - r_o)t^{-1/2}]$ are shown by dots and computation results according to the described model shown by dashed curves for the following materials: 1—ceramics, 2—teflon-b, 3—composite, 4—rubber, 5—gypsum, and 6—teflon-w. The dot shapes correspond to acquisition time. Inset outlines the experimental setup.

Material	Measured and standard values of a (mm^2/s)	Test number							
		1	2	3	4	5	6	7	8
Ceramics	a	1.041	1.010	1.033	1.091	1.076	1.046	1.001	1.040
	δa	0.011	0.018	0.017	0.026	0.012	0.015	0.019	0.010
	a_8	1.042 ± 0.016							
	a^*	1.120 ± 0.096							
Composite	a	0.238	0.244	0.237	0.240	0.227	0.233	0.237	0.238
	δa	0.014	0.019	0.011	0.044	0.010	0.009	0.017	0.005
	a_8	0.237 ± 0.016							
	a^*	0.276 ± 0.033							
Rubber	a	0.239	0.268	0.240	0.239	0.246	0.265	0.243	0.255
	δa	0.039	0.008	0.017	0.019	0.015	0.017	0.015	0.015
	a_8	0.249 ± 0.018							
	a^*	0.249 ± 0.032							
Teflon-w	a	0.119	0.114	0.115	0.129	0.127	0.113	0.123	0.128
	δa	0.006	0.005	0.004	0.005	0.007	0.004	0.007	0.007
	a_8	0.120 ± 0.005							
	a^{**}	0.118 ± 0.003							
Teflon-b	a	0.323	0.319	0.355	0.333	0.336	0.336	0.327	0.326
	δa	0.006	0.008	0.009	0.035	0.013	0.017	0.016	0.015
	a_8	0.332 ± 0.015							
	a^*	0.389 ± 0.055							
Gypsum	a	0.432	0.430	0.391	0.422	0.434	0.365	0.465	0.437
	δa	0.024	0.013	0.026	0.012	0.041	0.019	0.030	0.039
	a_8	0.422 ± 0.026							
	a^{**}	0.427 ± 0.007							

**Value obtained using standard HFM technique [28].*
***Standard reference value [74].*

Table 2.
Measured and reference values of a *for tested materials.*

Figure 5 shows elementary values of a; i.e., each dot corresponds to one frame of the film as a function of time t since the onset of heating for a number of tests of each material. As could be seen, the proposed technique and model produce very close values of a at a wide range of t depicted by oval curves. It allows collecting large statistics in one test to reduce measurement uncertainty originated from various random factors by averaging the set. Some variations in a value calculated based on different tests of the same material but with differing positions of heating center can be accounted for real inhomogeneity of material TPC.

To determine thermophysical properties of the same materials independently, standard steady-state HFM [28] for measuring thermal conductivity λ_{st} of ~1-mm thin plates with known passed heat flux has been used. Temperature diffusivity coefficient $a^* = \lambda_{st}/\rho c_p$ has been calculated for each material based upon independent measurements of density ρ and specific heat capacity c_p. As followed from **Table 2**, values of a, determined using the proposed technique, and the values of a^*, obtained

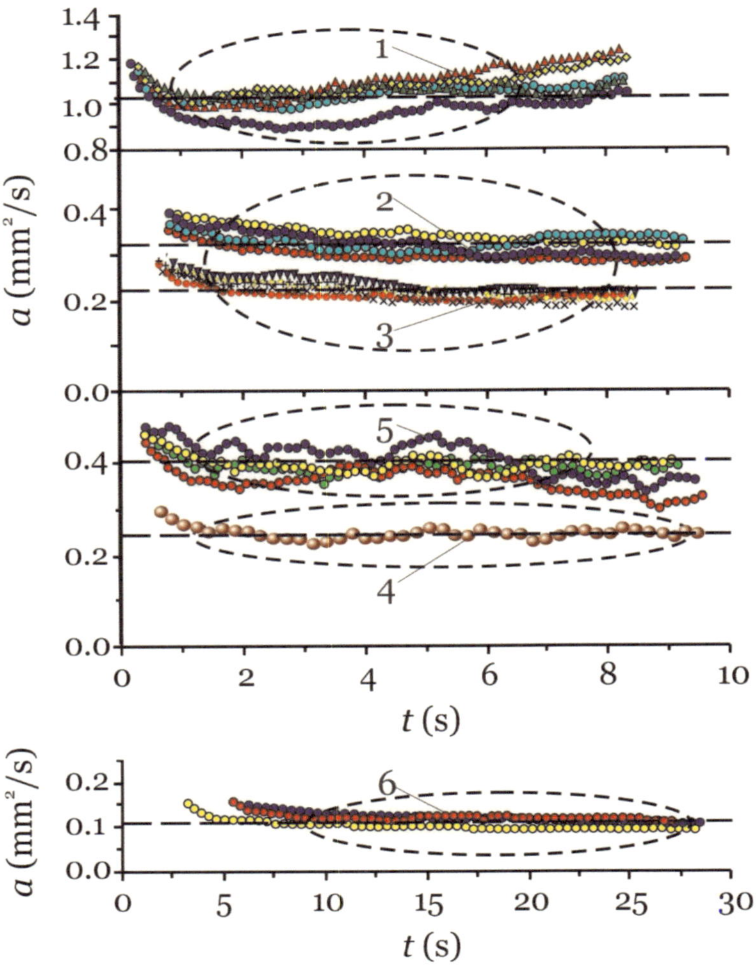

Figure 5.
Dependence of temperature diffusivity a on the time t passed since the heating onset. Dots correspond to experimental data and dashed lines show average value. 1—ceramics; 2—teflon-b; 3—composite; 4—rubber; 5—gypsum; and 6—teflon-w.

using standard steady-state technique, coincide within the stated inaccuracies of both the methods. For some of the materials, standard values a^{**} obtained from reference tables are shown instead of measured value a^{*}.

As could be seen from **Figure 5** and **Table 2**, relative random errors of a amount to several per cent. Systematic errors are generally of the same level. Origins of the latter ones are numerous, and the largest impact occurs from image scaling and optical aberrations of the camera, nonadiabatic heating growing with testing time, temperature dependence of TPC, especially in case of plastics, and transparency and reflectivity of studied material. Suppressing them is the objective of the next stage of the refinement technique.

As for effects of temperature dependence of TPC, they are determined by the following factors. Maximal temperature excess at the surface of a given material depends upon heat flux density and beam focusing radius. In all cases, it does not exceed a few tens of °C and several °C in the measurement area, so that the obtained results correspond to the temperature of environment in the first approximation. To reduce the possible impact of TPC temperature dependence on the measurement of a, one should either reduce temperature gradients in the measurement area or the area dimensions provided the gradients are the same. The former is achieved due to constant heating unlike instant heating in LFT. To reduce it further, which could be desirable for temperature-sensitive materials, one should increase heating spot radius r_0 and/or decrease energy density in an incident beam, but the former requires increasing measurement time τ, while the latter leads to reduction in signal/noise ratio, and both ways result in growing inaccuracies.

4. Thermal diffusivity measuring: 2D model

For thin plates, films or shells of highly conductive materials with thickness lying in the range $\delta \sim 0.1$–3 mm, it is possible to use similar technique, but forming cylinder thermal wave instead of spherical one. This approach has been widely used since the 1930s, it is standardized for measuring TPC of gaseous, liquid, solid, and noncoherent materials and is referred to as linear heat source or transient hot-wire (THW) technique [28, 75–79].

The proposed technique differs from the THW in heating zone geometry. In the standard technique, a cylinder thermal wave is induced by a linear heat source with the length much higher than thermal wave radius, while the proposed technique utilizes a point heat source at the surface of the plate, in which thickness is much less than the thermal wave radius. In both cases, temperature distribution is the same at distances several times higher than the maximal characteristic length of the problem, that is the highest of heating area size and plate thickness, and can be described by the following expression [73].

$$T(r,t) = (Q/4\pi\lambda\delta)\ln\left(4at/Cr^2\right) + O\left({r_0}^2/at\right) \tag{3}$$

where Q is the total power transferred to the plate, r_0 is the heating area radius, $\ln C = \gamma \approx 0.5772$ is the Euler constant, and $O(x)$ is some function limited by its argument multiplied at some finite constant. The second term in Eq. (3) may be neglected in practice for times high enough so that $t \gg \tau_F$. Unlike the case of a massive object and spherical thermal wave, the distribution $T(r, t)$ in this case does not approach to some steady asymptotic but keeps growing logarithmically with time. Provided heating power and testing time remain unchanged, the temperature reaches much higher values than in a massive object. It allows reaching significantly higher relative precision of TD measurements with the same power of laser beam or reducing the latter to decrease object overheating while keeping measurement accuracy.

Linear regression $y = d\cdot(x - g)$ in $(\ln(r\cdot t^{-1/2}), T)$ coordinates is used to calculate TD from the experimental data. As follows from Eq. (3), TD and the value of g are related as $a = \exp(2g - \gamma)/4$. Experiments were carried out on a low carbon steel plate with dimensions 100 × 100 × 2 mm. As a result of the processing of each frame of IR film, radial dependence $T(r)$ at different moments of time t since the heating onset has been found (see **Figures 2** and **6**) and then the values of a for each frame is found. Averaging a over several hundreds of frames of each film, one can obtain the mean value of a for each individual test. Statistical processing of 10 such mean values of a obtained in 10 different films results in $a = (12.54 \pm 0.27)$ mm^2/s (**Figure 7**). It coincides with the standard value of TD for low carbon steels, and relative root mean square of the set is as low as 2%. Using the above value of a, standard values for density $\rho = 7870$ kg/m^3 and specific heat capacity $c_p = 486$ J/(kg. K) for low carbon steel and relation $\lambda = a\rho c_p$, one can find the value of $\lambda = 48.0$ W/(m.K).

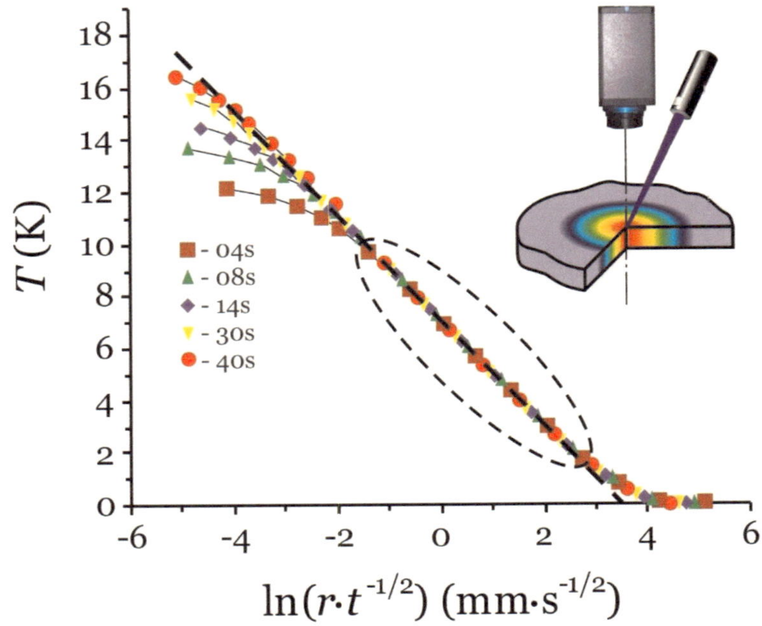

Figure 6.
Experimental data (dots) and their linear regression (dashed line) in 2D testing of a low carbon steel plate. Inset shows the experimental setup.

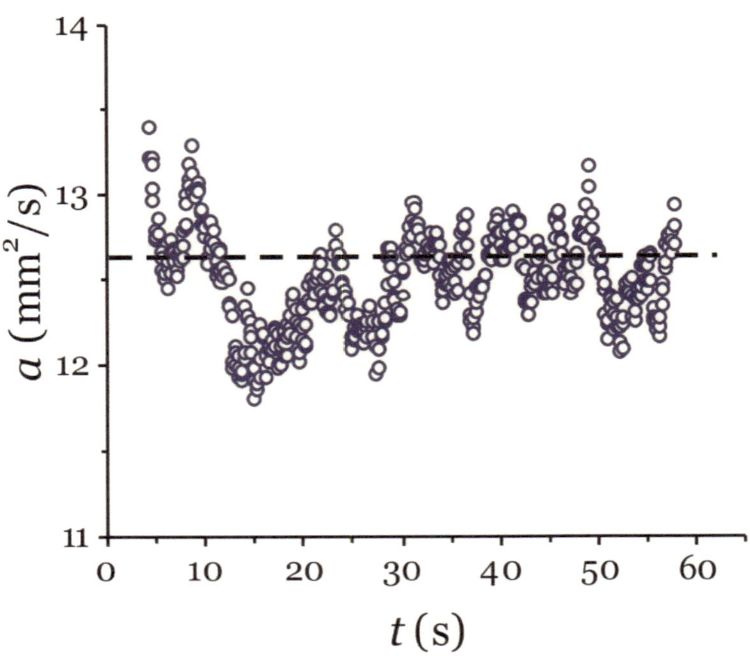

Figure 7.
Dependence of elementary values of temperature diffusivity a *over time* t *passed since the heating onset for a low carbon steel sample. Dots represent experimental results and dashed line shows an average value.*

5. Discussion of thermal diffusivity measuring

The two described variations of the presented technique complement each other. Spherical wave mode is applicable to any sample having dimension large enough (minimally $10r_0$), and testing duration is quite short, namely around a few tens of Fourier time for hot spot size. Typically, testing takes from a fraction of a second to a few seconds. Cylinder wave mode works better on highly conductive materials, but takes more time (usually tens of seconds) and imposes restrictions on sample thickness δ. From below, it is limited by growing impact of heat transfer through free surfaces, which is characterized by the ratio of the energy that left the sample through the surfaces to the stored energy at the column perpendicular to the plate plane $\alpha\tau/\rho c_p\delta$, where α is the total surface heat transfer coefficient including both convective and radiant parts for both surfaces. Increase in δ leads to proportionally growing external heat source power required to heat the sample to the same temperature. It also requires testing time proportional to thickness squared for thermal wave to travel minimally required distance $\sim 10\delta$ to achieve the measurement accuracy of a few per cents employing the proposed cylinder wave model. It should be mentioned that for plates with $\delta \sim 1$ mm, both cylinder and spherical models may be used, though the former requires much higher testing time τ, while the latter requires higher spatial and temporal resolution from the camera and much better laser beam focusing.

Let us stress one more time that both described modes of the proposed approach do not require absolute values of surface temperature or heat fluxes, whereas neither data processing algorithms nor equation for determining a value do not require them. The only assumption of IR radiation value uniformity across the surface is enough. Possible optical inhomogeneities are suppressed significantly at the stage of pixel-by-pixel subtraction of the base image obtained before heating onset from every following image.

So that, the paper presents new express nondestructive testing technique for TPC measurement for both massive objects and thin plates including bulky components and elements of finished products, which does not require extensive sample preparation. Testing the technique ascertains that TD coefficient measured according to the proposed procedure coincides with known standard value or value measured by independent steady-state method with good accuracy.

The described approach can be considered as a further development of standard laser flash technique [14–17]. Low standard errors and low requirements for measurement conditions are mostly due to analysis of not just the only function $T(t)$ of one argument acquired just once per testing, but large two-dimensional array of data $T(r, t)$, while for each given value of t, the dependence $T(r)$ results from the averaging of tens to hundreds of points located at the same distance to the center of the heating spot. Recording spatial and temporal dynamics of thermal wave propagation allows averaging temperature data twice—firstly at r = const over the angle φ up to 0–360° range averaging several tens or even hundreds of pixels, and then over time by data analysis of several hundreds of individual frames of IR film.

In case of anisotropic materials of macroscopic inhomogeneities, averaging temperatures from pixels located at r = const can be done at $\varphi < 360^\circ$ and then analyzed and plotted as $a(\varphi)$ or processing as usual just excluding the arcs where discrepancies exceed noise level by a given relative value. In the latter case, it can be an indication of material defects and could be a basis for developing highly sensitive differential techniques of thermographic nondestructive testing [68]. The region selected this way can be inspected in more detail on demand by changing the heating center position, or results of several tests for TD measurement using φ = 360° can be averaged to diminish inaccuracies further.

The main sources of systematic errors of the proposed technique are:

1. Thermographic techniques are affected by the scaling factor linking pixels and real length; for this reason, special care should be taken in the procedures for

determining actual geometrical factor. Additional inaccuracies can be brought about by errors in frame timestamps written by camera or personal computer running nonrealtime operating system.

2. Discrepancy with the model due to:

   a. incomplete suppression of optical inhomogeneities of the surface by image processing software,

   b. nonpoint-like heating especially in semitransparent material,

   c. stray and reflected light and IR-radiation, including that originating from laser beam collimation system; and

   d. imperfect abiding to adiabatic condition, i.e., significant convective or radiation heat flow across the surface.

6. Nondestructive testing using point heating

Thermographic approaches to nondestructive testing (T-NDT) seem promising for a wide variety of specific applications [3–7]. They can be multifunctional, multiscale, and contactless [20, 24, 35, 79–84]. Pulsed techniques have a number of advantages over quasi-stationary ones [20, 24, 69, 79, 80, 85] such as higher information content, higher spatial resolution, higher sensitivity to inhomogeneities, on the other hand, they require more powerful heat sources, higher resolution of data acquisition system, and more sophisticated thermal-physical models and data processing algorithms. It is convenient to create transient thermal fields by short-term heat supply to the area of interest. There are a lot of T-NDT techniques utilizing IR-cameras capable to resolve numerous issues concerning defect or suspicious sites detection and localization [6–8]. However, its potential has not been fully realized yet, and further advances in IR-cameras spatial and temporal resolution and data processing algorithms enable resolving more and more complex problems in monitoring and diagnostics.

We further describe the experimental and data processing techniques combining crack detection, localization, and characterization with enhanced signal-to-noise ratio and sensitivity to small defects.

6.1 Stepwise point heating technique for NDT

The technique is based upon computer analysis of transient thermal field generated by an external heat source localized at the spot with 0.1–0.3 mm radius [85]. Such a source produces radially symmetric heat wave in a defect-free plate, so that the position of concentric isotherms at any time can be obtained with high accuracy by averaging signals from a large number of IR-camera pixels at r = const as described in the first part of the paper. A defect locally contorts the isotherm circular shape, making it a good diagnostic indicator. The technique is differential at the root that provides high sensitivity and signal/noise ratio of the testing.

A number of model defects, namely part-through notches made on either front or rear sides of the metal plates by a thin (0.8 mm) abrasive microdisk, have been studied experimentally. These tests give temperature distributions related to typical defects and allow evaluation of the developed model and software adequacy for thermal image analysis. To reduce camera matrix random noise, the signal from every pixel has been averaged over 10–60 sequential frames. The experiments have been carried out on the plate with 3-mm thickness made of mild carbon steel "Steel 40X" (analog of mild carbon steel AISI 5140).

6.2 NDT results

As can be seen in **Figure 1** presenting typical raw data, experimentally obtained isotherms at a defect-free plate can be approximated with concentric circles with high accuracy. A defect distorts circular shape of isotherms locally. Thus, **Figure 8** demonstrates an instance of the temperature field in an area containing part-through notch with length L = 15 mm and depth h = 2.5 mm at the rear side of the plate. Isotherms distortion induced by the notch is quite specific. They swell outward from the center at the cut bank facing the heating center and inward at the opposite bank (see **Figure 8b**). The above distortion is due to the additional thermal resistance introduced by the notch to the thermal flow, so that the temperature at the notch side facing the heat source goes slightly up relative to the defect-free area at the same distance from the heat source, while the temperature at the other side goes slightly down.

The data discussed above comprise an input for further processing. Subtraction of temperature TR averaged over angle φ over the defect-free region from the T allows detecting temperature field variations ΔT_d, originating from much smaller defects than could be seen by the naked eye. **Figure 9** presents distributions of differential temperature ΔT_d at the surface in two perpendicular directions along and across the plane of the part-through notch with length L = 15 mm and depth h = 2.5 mm on the hidden side of the plate. The maximal temperature difference ΔT_{max} between the points with the highest and the lowest ΔT_d in the region around the notch depends on its length, depth, angle φ between its axis and direction to the heating center (**Figure 10a**) and distance to it (**Figure 10b**). At the most favorable circumstances, this technique is capable to detect and characterize notches with length L around a few millimeters at the signal to noise ratio 3:1. Weak ΔT_{max} dependence on angle φ provided $\varphi > 10°$ and rather wide range of distances R between the defect and the heating center where the distance dependence is also weak allows one to detect and characterize multiple cracks with unknown position and orientation in a certain region in a single test.

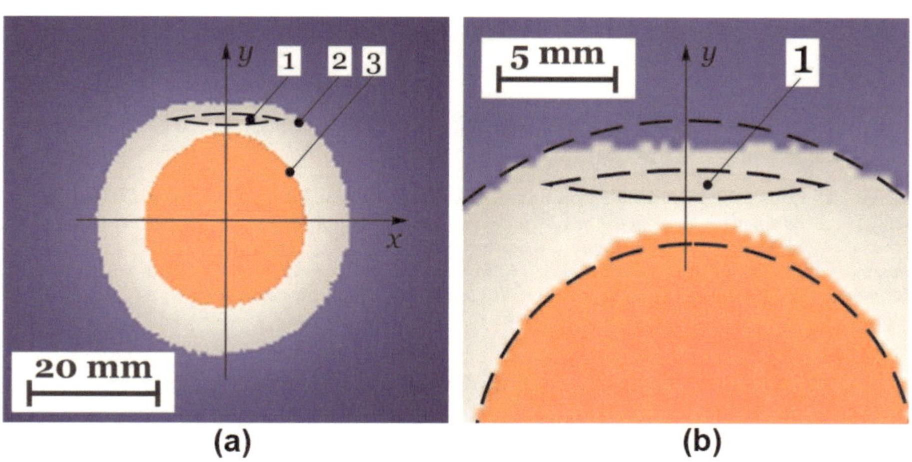

Figure 8.
Typical temperature distribution resulting from local heating of the steel plate containing part-through notch with L = 15 mm and depth h = 2.5 mm during 60 s in different scales (a) and (b) with superimposed isotherms. 1—notch, 2—T = 15°C, 3—T = 20°C.

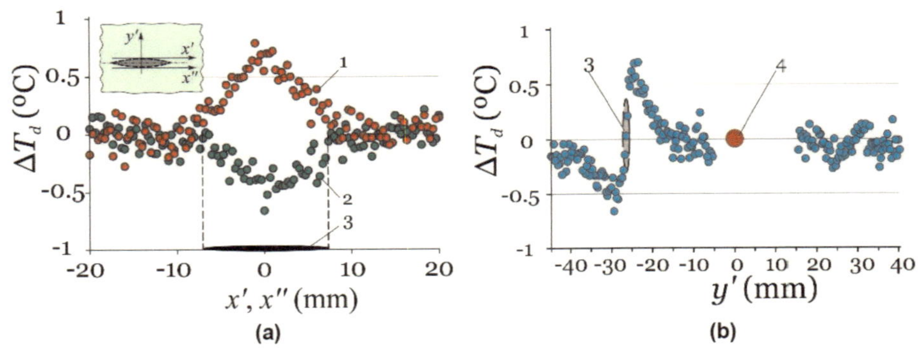

Figure 9.
Temperature profiles along x′ and x axes (a) and along y′ axis (b) after 60 s of local heating at zero point of the steel plate containing 15-mm-length part-through notch at the hidden side. 1—in front of the notch, 2—behind the notch, 3—notch position, 4—heating center, and ΔT_d—temperature difference.

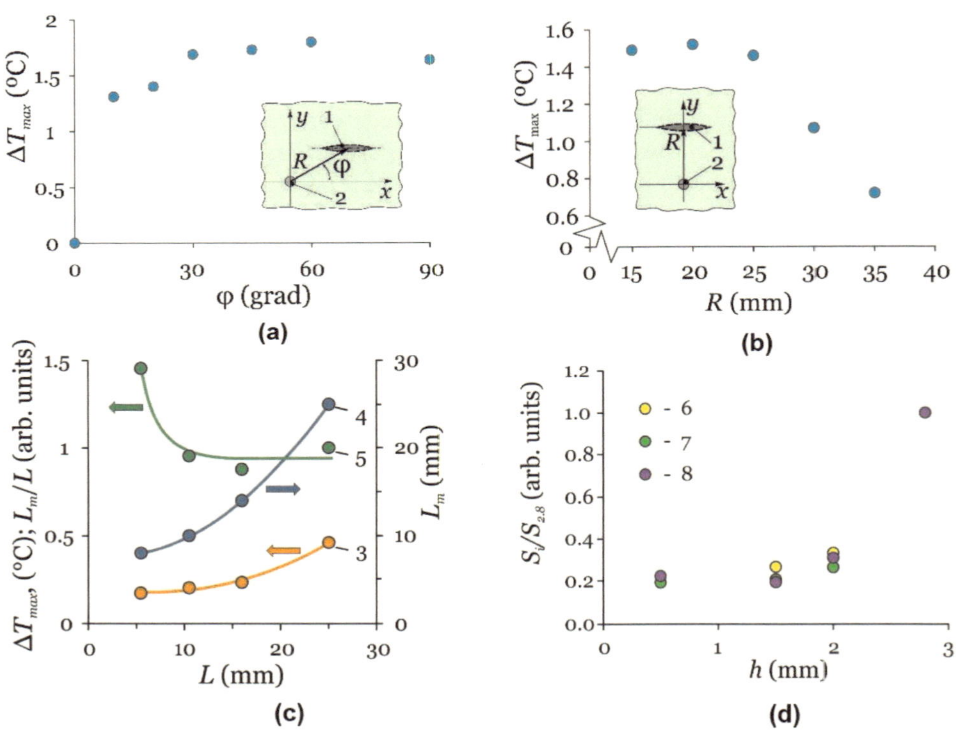

Figure 10.
Experimentally found dependences of values related to part-through notches on the hidden side of the plate. (a) ΔT_{max} over the angle φ between the notch plane and radius-vector R from the heating center to the notch center as shown in the inset (1—notch and 2—heating center); (b) ΔT_{max} over the distance R; (c) 3—ΔT_{max}, 4—width of ΔT_d profile raise L_m, and 5—L_m/L ratio over the notch length L; (d) signal ratio $S_i(h)/S_i(2.8)$ at different data processing iterations (6—S_0; 7—S_3; and 8—S_{4d}) over the notch depth h. ΔT_{max} is maximal temperature difference at the region near the notch.

The width of the raise in ΔT_d distribution along the notch is well correlated with the notch length L (see **Figure 10c**). So, it is a convenient value to characterize the length L, and the experiment results (see **Figure 10c**) prove that starting from several millimeters, the length L can be determined with the accuracy better than 10%. The maximum of the distribution ΔT_{max} weakly depends upon the notch length and cannot be used as a reliable indicator. ΔT_{max} dependence upon the notch depth h is prominently nonlinear, and for shallow notches with depth up to 2/3 of plate thickness, it is quite weak too (see **Figure 10d**). It does not allow reliable determination of the depth of such defects.

6.3 Shallow notch detection enhancing

The signal originated from the notch is obscured by noise for $h < 0.5\,\delta_m$ where δ_m is the plate thickness, so that it is very difficult not only to evaluate the notch depth from raw signal, but even to detect its presence. To detect such defects producing low-amplitude short-range deviations of a specific pattern in the temperature field, it is possible to use gradient methods with a large window with the size smaller but comparable to the characteristic defect fingerprint extension. So we propose the following procedure: for every pixel, the absolute value of temperature gradient is calculated at quite a large window like 9 × 9 or 13 × 13 pixels that is about 3 × 3 or 4 × 4 mm, producing a slope map of the original image. It reduces noise drastically and enhances field irregularities with the characteristic size about that of the window. This operation is then repeated several times in diminishing windows, so that the slope image of the slope image, etc. is calculated, which are further referred to as $S_{i+1} = |\text{grad}\ S_i|$ where i is the iteration number. Successive iterations do not significantly affect the ratio of signals produced by the notches of different depth as shown in **Table 3** and **Figure 10d**, but notably improve signal/noise ratio, thus enhancing our capability to detect shallow notch as shown in **Table 4** and **Figure 11**. Additional way to improve signal/noise ratio is to obtain the differences $S_{id} = \partial S_i/\partial x$ between adjacent points of S_i along the line perpendicular to the notch. Thus, preliminary detection technique can be based upon a local exceedance of some threshold value of S_i, followed by confirmation using S_{id} level.

The result of four successive computations of gradient absolute values S_1–S_4 over the full image is shown in **Figure 11** at the lines perpendicular to the notches along with S_{4d}, which is a signed uniaxial derivative of S_4 along the selected line. As could be seen, the notch with h = 0.5 mm located at x = 20 mm can be reliably detected by signal level from the third iteration on. **Table 4** presents signals from notches with varying depths normalized to the signal of notch with h = 2.8 mm, at various processing stages. S_0 is the raw signal averaged in 5 × 5 pixel window. **Table 2** presents signal/noise ratios for two notches with length L = 15 mm and depths of 0.5

***h* (mm)**	**Iteration**							
	S_0	S_1	S_2	S_{2d}	S_3	S_{3d}	S_4	S_{4d}
0.5	0.22	0.20	0.19	0.24	0.19	0.22	0.22	0.22
1.5	0.26	0.24	0.20	0.30	0.20	0.23	0.20	0.19
2.0	0.33	0.29	0.27	0.32	0.26	0.35	0.28	0.31

Table 3.
Dependence of sensitivity to crack depth h *as* $S_i(h)/S_i(2.8)$ *upon signal processing method.*

***h* (mm)**	**Iteration**							
	S_0	S_1	S_2	S_{2d}	S_3	S_{3d}	S_4	S_{4d}
0.5	0.51	1.16	1.69	2.15	1.88	2.38	2.40	3.10
2.8	2.36	5.77	9.03	8.93	9.76	10.6	11.0	14.0

Table 4.
Dependence of signal/noise ratio for two cracks upon signal processing method.

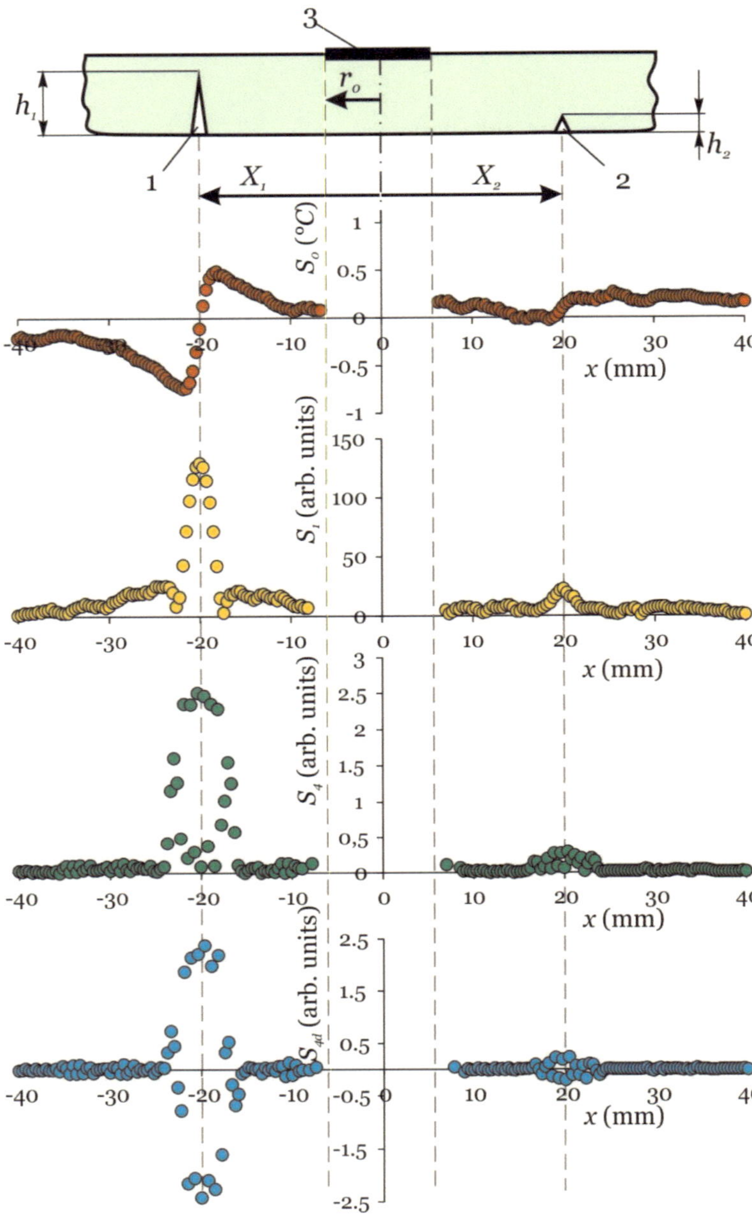

Figure 11.
Profiles of the signal at different stages of data processing along the x axis. S_0 is raw temperature T averaged over 1.5 × 1.5 mm² area, i.e., 5 × 5 pixels. The heating region is marked by the dashed lines. $S_1 = |grad\ T|$, $S_4 = |grad\ S_3|$, $S_{4d} = \partial S_4/\partial x$. Inset schematically presents experiment layout. Notches h_1 (1) and h_2 (2) ($h_1 > h_2$) and heating region (3).

and 2.8 mm at the same processing stages. Both signal and noise values in the tables refer to the peak values.

7. Conclusions

Let us briefly summarize the most important characteristic features and main advantages of the proposed technique justified by the experimental results:

1. The proposed technique is adequate within a wide time range providing several hundreds of individual values of a differing by no more than 10–15% to obtain the averaged result from the film. Moreover, each individual measurement results from processing temperature distribution obtained, in turn, by averaging large number of pixels located at the same distance from the heating center. Such double averaging results in very low random error, which is around a few per cent and supplies rich statistics providing quite accurate measurements. The best results were obtained on testing stabilized zirconia ceramics, which has the most uniform optical properties and microstructure of all tested materials. Relative random error of TD measured in single test was below 2% for this material.

2. Short duration τ of the testing—for the tested materials with moderate values of $a \sim 0.1$–15 mm^2/s, τ = 3–20 s. τ decreases with growing a and better heating localization.

3. Continuous heating during several seconds of testing as opposed to "instant" heating by very high peak intensity pulse in LFT excludes severe overheating of sample material. At any time and for all materials, maximal overheatings in the heating spot and in the measurement area do not exceed several tens and several kelvins accordingly. For the vast majority of cases, it allows neglecting the effect of temperature dependence of temperature diffusivity on obtained value.

4. Consequently, low duration τ leads to diminishing inaccuracies originated from heat losses across the surfaces, contributions from surrounding constructions, drifts, and others.

5. Loose requirements for sample preprocessing and geometry—minimal 3D sample dimensions are 5–10 mm in any direction from the heating center, and the surface in this area should be macroscopically flat, namely surface roughness should not exceed 20–50 mkm. Higher roughness values result in increasing measurement random errors, but short testing duration makes statistics collection easy. Corrections for surface curvature can be introduced if necessary.

6. Weak dependence or almost complete independence of surface optical properties impact on the measurement results apart from transparent materials with low factors of volume absorption and stray light.

7. Good results reproducibility with TD root mean square in order of a few per cent.

8. Testing is still possible if only one side of the tested object is freely available.

9. Technical plausibility of designing inexpensive portable device for express testing various objects including large sized ones. Due to its simplicity and high testing rate, it can be installed in production lines for *in situ* quality control or used for out-of-the-laboratory control of finished products quality.

10. Dynamic thermographic technique has been developed for testing shells, plates, vessels, tubes, reactors, fuselage elements, etc. It is capable of detecting defects with length above a few millimeters and depth more than a

few tenths of millimeters not only at the visible but also at the inner unreachable side of the shell.

Acknowledgements

The work was carried out with financial support from the Ministry of Education and Science of the Russian Federation in the framework of Russian Science Foundation project no. 15-19-00181 (carrying out experiment and data processing), Russian Foundation for Basic Research grant no. 17-48-680817 (development of analytical and computational models), and Ministry of Education and Science of the Russian Federation project no. 16.2100.2017/4.6 (building and debugging of original pilot installation). Experimental results have been obtained using equipment of Core Facilities Center of Derzhavin Tambov State University.

Conflicts of interest

The authors declare no conflict of interest.

Author details

Dmitry Yu. Golovin[1], Alexander G. Divin[2], Alexander A. Samodurov[1], Alexander I. Tyurin[1] and Yuri I. Golovin[1,3*]

1 Research Institute for Nanotechnologies and Nanomaterials, Derzhavin Tambov State University, Tambov, Russia

2 Tambov State Technical University, Tambov, Russia

3 Chemical Department, Lomonosov Moscow State University, Moscow, Russia

*Address all correspondence to: yugolovin@yandex.ru

References

[1] Vavilov VP. Thermal NDT: Historical milestones, state-of-the-art and trends. Quantitative InfraRed Thermography Journal. 2014;**11**:66-83

[2] Hudson RD. Infrared System Engineering. New York (NY): Wiley-Interscience; 1969. p. 672

[3] Haddadnia J, Zadeh HG. The New Trends in the Application of Thermography Science for Diagnostic Purposes. Reseda, CA, USA: Supreme Century; 2016. p. 152

[4] Gaussorgues G, Chomet S. Infrared Thermography. Netherlands: Springer Science and Business Media; 2012. p. 508

[5] Vollmer M, Möllmann K-P. Infrared Thermal Imaging: Fundamentals, Research and Applications. Wiley-VCH: Weinheim; 2010. p. 612

[6] Meola C, Ibarra-Castanedo C, Maldague XPV. Infrared Thermography Recent Advances and Future Trends (Bentham eBooks). 2012Infrared thermography. In: Czichos H, editor. Handbook of Technical Diagnostics. Berlin, Heidelberg: Springer-Verlag; 2013. pp. 175-220

[7] Usamentiaga R, Venegas P, Guerediaga J, Vega L, Molleda J, Bulnes FG. Infrared thermography for temperature measurement and non-destructive testing. Sensors. 2014;**14**: 12305-12348

[8] Ciampa F, Mahmoodi P, Pinto F, Meo M. Recent advances in active infrared thermography for non-destructive testing of aerospace components. Sensors. 2018; **18**:609-646

[9] Prakash RM. Infrared Thermography. Rijeka, Croatia: InTech; 2012. p. 245

[10] Breitenstein O, Warta W, Langenkamp M. Lock-in Thermography: Basics and Use for Evaluating Electronic Devices and Materials. New York: Springer; 2010. p. 257

[11] Bar-Cohen A, Matin K, Narumanchi S. Nanothermal interface materials: Technology review and recent results. Journal of Electronic Packaging. 2015; **137**:040803-1-040803-17

[12] Choudhary A, Goyal D, Shimy SL, Akula A. Condition monitoring and fault diagnosis of induction motors: A review. Archives of Computational Methods in Engineering. 2018:1-18. Available from: https://doi.org/10.1007/s11831-018-9286-z

[13] Xingwang G, Vavilov VP. Pulsed thermographic evaluation of disbonds in the insulation of solid rocket motors made of elastomers. Polymer Testing. 2015;**45**:31-40

[14] Subramainam B, Lahiri BB, Saravanan T, Philip J. Infrared thermography for condition monitoring —A review. Infrared Physics and Technology. 2013;**60**:35-55

[15] Zhang H, Yang R, He Y, Foudazi A, Cheng L, Tian G. A review of microwave thermography nondestructive testing and evaluation. Sensors. 2017;**17**:1123-1156

[16] Kordatos EZ, Exarchos DA, Stavrakos C, Moropoulou A, Matikas TE. Infrared thermographic inspection of murals and characterization of degradation in historic monuments. Construction and Building Materials. 2013;**48**:1261-1265

[17] Theodorakeas P, Cheilakou E, Ftikou E, Koui M. Passive and active infrared thermography: An overview of applications for the inspection of mosaic

structures. Journal of Physics: Conference Series. 2015;**655**:012061

[18] Garrido I, Lagüela S, Arias P. Infrared thermography's application to infrastructure inspections. Infrastructures. 2018;**35**:1-19

[19] Lucchi E. Applications of the infrared thermography in the energy audit of buildings: A review. Renewable and Sustainable Energy Reviews. 2018; **82**:3077-3090

[20] Vavilov VP, Burleigh DD. Review of pulsed thermal NDT: Physical principles, theory and data processing. NDT&E International. 2015; **73**:28-52

[21] Balageas D, Maldague XD, Burleigh D, Vavilov VP, Oswald-Tranta JM, Roche B, et al. Thermal (IR) and other NDT techniques for improved material inspection. Journal of Nondestructive Evaluation. 2016;**35**:1-17

[22] Ibarra-Castanedo C, Tarpani JR, Maldague XPV. Nondestructive testing with thermography. European Journal of Physics. 2013;**34**:S91-S109

[23] Vavilov VP. Thermal nondestructive testing of materials and products: A review. Russian Journal of Nondestructive Testing. 2017;**53**: 707-730

[24] Vavilov VP. Dynamic thermal tomography: Recent improvements and applications. NDT&E International. 2015;**71**:23-32

[25] Yang R, He Y. Optically and non-optically excited thermography for composites: A review. Infrared Physics and Technology. 2016;**75**:26-50

[26] Zhao Y, Addepalli S, Sirikham A, Roy R. A confidence map based damage assessment approach using pulsed thermographic inspection. NDT and E International. 2018;**93**:86-97

[27] Bodnar JL. Crack detection by stimulated infrared thermography. European Physical Journal Applied Physics. 2014;**65**:31001-31012

[28] Yüksel N. Chapter 6: Insulation materials in context of sustainability. In: Almusaed A, Almssad A, editors. The Review of Some Commonly Used Methods and Techniques to Measure the Thermal Conductivity of Insulation Materials. InTech; 2016. pp. 113-140

[29] Cernuschi F, Bison PG, Figari A, Marinetti S, Grinzato E. Thermal diffusivity measurements by photothermal and thermographic techniques. International Journal of Thermophysics. 2004;**25**:439-457

[30] Abad B, Borca-Tasciuc D-A, Martin-Gonzalez MS. Non-contact methods for thermal properties measurement. Renewable and Sustainable Energy Reviews. 2017;**76**: 1348-1370

[31] Hammerschmidt U, Hameury J, Strnad R, Turzó-Andras E, Wu J. Critical review of industrial techniques for thermal-conductivity measurements of thermal insulation materials. International Journal of Thermophysics. 2015;**36**:1530-1544

[32] Doshvarpassand S, Wu C, Wang X. An overview of corrosion defect characterization using active infrared thermography. Infrared Physics and Technology. 2019;**96**:366-389

[33] Vostikolaei FS, Akhoondzadeh M. A comparison of four methods for extracting land surface emissivity and temperature in the thermal infrared hyperspectral data. Earth Observation and Geomatics Engineering. 2018;**2**: 56-63

[34] Pereira CB, Yu X, Dahlmanns S, Blazek V, Leonhardt S, Teichmann D. Chapter 1: Infrared thermography. In: Abreu de Souza M et al., editors. Multi-

Modality Imaging. Switzerland AG: Springer Nature; 2018. pp. 1-29

[35] Lahiri BB, Bagavathiappan S, Jayakumar T, Philip J. Medical applications of infrared thermography: A review. Infrared Physics and Technology. 2012;**55**:221-235

[36] JIP Q. Application of Infrared Thermography in Sports Science. Switzerland: Springer; 2017. p. 339

[37] Zhao D, Qian X, Gu X, Jajja SA, Yang R. Measurement techniques for thermal conductivity and interfacial thermal conductance of bulk and thin film materials. Journal of Electronic Packaging. 2016;**138**:040802-1-040802-19

[38] Okamoto Y, Watanabe S, Ogata K, Hiramatsu K, Miyazaki H, Morimoto J. Proposal of novel measurement method for thermal diffusivity from infrared thermal movie. Japanese Journal of Applied Physics. 2017;**56**:056601

[39] Tailor RE, Maglik KD. Pulse method for thermal diffusivity measurement, compendium of thermophysical property measurement methods. In: Maglic KD et al., editors. Survey of Measurement Techniques. Vol. 1. New York: Plenum Press; 1984. pp. 305-336

[40] Chernuschi F, Figari A, Fabbri L. Thermal wave interferometry for measuring the thermal diffusivity of thin slabs. Journal of Materials Science. 2000;**35**:5891-5897

[41] Parker WJ, Jenkins RJ, Butler CP, Abbot GL. Flash method of determining thermal diffusivity, heat capacity and thermal conductivity. Journal of Applied Physics. 1961;**32**:1679-1684

[42] Vozar L, Hohenauer W. Flash method of measuring the thermal diffusivity. A review. High Temperatures - High Pressures. 2003/2004;**35/36**:253-264

[43] McMasters RL, Dinwiddie RB. Anisotropic thermal diffusivity measurement using the flash method. Journal of Thermophysics and Heat Transfer. 2014;**28**:518-523

[44] Kruczek T, Adamczyk WP, Bialecki RA. In situ measurement of thermal diffusivity in anisotropic media. International Journal of Thermophysics. 2013;**34**:467-485

[45] Wang L, Gandorfer M, Selvam T, Schwieger W. Determination of faujasite-type zeolite thermal conductivity from measurements on porous composites by laser flash method. Materials Letters. 2018;**221**:322-325

[46] Salazar A, Mendioroz A, Apiñaniz E, Pradere C, Noël F, Batsale J-C. Extending the flash method to measure the thermal diffusivity of semitransparent solids. Measurement Science and Technology. 2014;**25**:035604

[47] Coquard R, Panel B. Adaptation of the flash method to the measurement of the thermal conductivity of liquids or pasty materials. International Journal of Thermal Sciences. 2009;**48**:747-760

[48] Shao X, Mo S, Chen Y, Yin T, Yang Z, Jia L, et al. Solidification behavior of hybrid TiO_2 nanofluids containing nanotubes and nanoplatelets for cold thermal energy storage. Applied Thermal Engineering. 2017;**117**:427-436

[49] Noroozi M, Zakaria A. Chapter 4: Measuring nanofluid thermal diffusivity and thermal effusivity: The reliability of the photopyroelectric technique. In: Nanofluid Heat and Mass Transfer in Engineering Problems. Intech; 2017. pp. 65-93

[50] Azmi W, Sharma K, Mamat R, Najafi G, Mohamad M. The enhancement of effective thermal conductivity and effective dynamic

viscosity of nanofluids—A review. Renewable and Sustainable Energy Reviews. 2016;**53**:1046-1058

[51] Murshed S, De Castro C. Superior thermal features of carbon nanotubesbased nanofluids—A review. Renewable and Sustainable Energy Reviews. 2014;**37**:155-167

[52] Tertsinidou G, Tsolakidou C, Pantzali M, Assael M. New measurements of the apparent thermal conductivity of nanofluids and investigation of their heat transfer capabilities. Journal of Chemical and Engineering Data. 2017;**62**:491-507

[53] Laser Flash Analysis—LFA. Method, Technique, Applications. NETZSCH-Geratebau GmbH [Internet]. 2019. Available from: http://www.netzsch.com [Accessed: 11 June 2019]

[54] Discovery Flash. [Internet]. 2019. Available from: https://www.tainstruments.com [Accessed: 11 June 2019]

[55] ISO 22007-4. Plastics—Determination of thermal conductivity and thermal diffusivity—Part 4: Laser flash method; 2017

[56] ISO 18755. Fine ceramics (advanced ceramics, advanced technical ceramics). Determination of thermal diffusivity of monolithic ceramics by laser flash method; 2005

[57] ASTM E1461-13. Standard test method for thermal diffusivity by the flash method

[58] Donaldson AB, Taylor RE. Thermal diffusivity measurement by a radial heat flow method. Journal of Applied Physics. 1975;**46**:4584-4589

[59] Cernuschi F, Russo A, Lorenzoni L, Figari A. In-plane thermal diffusivity evaluation by infrared thermography. Review of Scientific Instruments. 2001; **72**:3988-3995

[60] Dong H, Zheng B, Chen F. Infrared sequence transformation technique for in situ measurement of thermal diffusivity and monitoring of thermal diffusion. Infrared Physics and Technology. 2015;**73**:130-140

[61] Genna S, Ucciardello N. A thermographic technique for in-plane thermal diffusivity measurement of electroplated coatings. Optics and Laser Technology. 2019;**113**:338-344

[62] Bedoya A, González J, Rodríguez-Aseguinolaza J, Mendioroz A, Sommier A, Batsale JC, et al. Measurement of in-plane thermal diffusivity of solids moving at constant velocity using laser spot infrared thermography. Measurement. 2019;**134**:519-526

[63] Cielo P, Utracki LA, Lamontagne M. Thermal diffusivity measurements by the converging-thermal-wave technique. Canadian Journal of Physics. 1986;**64**:1172-1177

[64] Murphy F, Kehoe T, Pietralla M, Winfield R, Floyd L. Development of an algorithm to extract thermal diffusivity for the radial converging wave technique. International Journal of Heat and Mass Transfer. 2005;**48**:395-402

[65] Kehoe T, Murphy F, Kelly P. A method for measuring the thermal diffusivity of intermediate thickness surface absorbing samples and obtaining the ratio of anisotropy by the converging wave flash method. International Journal of Thermophysics. 2009;**30**:987-1000

[66] Kim SW, Kim JC, Lee SH. Analysis of thermal diffusivity by parameter estimation in converging thermal-wave technique. International Journal of Heat and Mass Transfer. 2006;**49**:611-616

[67] Husin MS, Moksin MM, Jibrin S, Azmi BZ, Yunus WMM, Waziri M. A simplified low cost converging thermal wave technique for measuring thermal

diffusivity of thin foils. IOSR Journal of Applied Physics. 2013;**3**:38-46

[68] Golovin YI, Turin AI, Golovin DY, Samodurov AA. New methods of thermographic control using multi-scale analysis of non-stationary thermal fields. Industrial Laboratory (Diagnostics of Materials). 2018;**84**: 23-33

[69] Almond DP, Angioni SL, Pickering SG. Long pulse excitation thermographic non-destructive evaluation. NDT&E International. 2017; **87**:7-14

[70] Badghaish AA, Fleming DC. Non-destructive inspection of composites using step heating thermography. Journal of Composite Materials. 2008; **42**:1337-1357

[71] Roche J-M, Balageas D. Common tools for quantitative pulse and step-heating thermography—Parts l and ll. QIRT Conference; Bordeaux, France; 2014. pp. 1-28

[72] Palumbo D, Cavallo P, Galietti U. An investigation of the stepped thermography technique for defects evaluation in GFRP materials. NDT and E International. 2019;**102**: 254-263

[73] Carslaw HC, Jaeger JC. Conduction of Heat in Solids. Vol. 510. USA: Oxford University Press; 1959

[74] Babichev AP, Babushkina NA, Bratkovskii AM, et al. In: Grigoriev IS, Melikhova EZ, editors. Handbook on Physical Quantities. Moscow: EnergoAtomIzdat; 1991. p. 1232. (in Russian)

[75] Watanabe H. Further examination of the transient hot-wire method for the simultaneous measurement of thermal conductivity and thermal diffusivity. Metrologia. 2002; **39**:65-81

[76] Merckx B, Dudoignon P, Garnier JP, Marchand D. Simplified transient hot-wire method for effective thermal conductivity measurement in geo materials: Microstructure and saturation effect. Advances in Civil Engineering. 2012;**2012**:1-10

[77] Assael MJ, Dix M, Gialou K, Vozar L, Wakeham WA. Application of the transient hot-wire technique to the measurement of the thermal conductivity of solids. International Journal of Thermophysics. 2002;**23**: 615-633

[78] Guo W, Li G, Zheng Y, Dong C. Measurement of the thermal conductivity of SiO_2 nanofluids with an optimized transient hot wire method. Thermochimica Acta. 2018;**661**:84-97

[79] Assael MJ, Antoniadis KD, Wakeham WA. Historical evolution of the transient hot-wire technique. International Journal of Thermophysics. 2010;**31**:1051-1072

[80] Liang K, Ma Y, Xie Y, Zhou B, Wang R. A new adaptive contrast enhancement algorithm for infrared images based on double plateaus histogram equalization. Infrared Physics and Technology. 2012;**55**:309-315

[81] Gao B, Woo WL, Tian GY. Electromagnetic thermography nondestructive evaluation: Physics-based modeling and pattern mining. Scientific Reports. 2016;**6**:25480

[82] Qiu J, Pei C, Liu H, Chen Z, Demachi K. Remote inspection of surface cracks in metallic structures with fiber-guided laser array spots thermography. NDT and E International. 2017;**92**:213-220

[83] Xu C, Gong X, Xumei X, Zhang W, Chen G. An investigation on eddy current pulsed thermography to detect surface cracks on the tungsten carbide matrix of polycrystalline diamond

compact bit. Applied Sciences. 2017;**7**: 1-12

[84] Tavakolian P, Sivagurunathan K, Mandelis A. Enhanced truncated-correlation photothermal coherence tomography with application to deep subsurface defect imaging and 3-dimensional reconstructions. Journal of Applied Physics, 2017;**122**:023103–1–023103–10

[85] Golovin YI, Tyurin AA, Golovin DY, Samodurov AA. Radially-symmetric thermal wave as a source of information on the state of ingomogenity in plates and sheaths. Russian Physics Journal. 2018;**61**:1472-1477

Chapter 8

Evaluation of Steel Rebar in Concrete Using Electromagnetic Method

Dongfeng He

Abstract

The corrosion of steel reinforcing bar (rebar) is the leading cause of deterioration of concrete. In Japan, many railway bridges were built 40 years ago. It is necessary to develop easy-operation method to evaluate the corrosion of steel rebar. A project about the corrosion evaluation of steel rebar was started in 2015. In this project, we have two objectives: one is to evaluate the depth and the diameter of steel rebar in concrete; another is to evaluate the corrosion of steel rebar in concrete. We developed electromagnetic methods to do nondestructive evaluation of the steel reinforcing bar (rebar) in concrete. Using two probes and lower excitation frequencies of 3.8 and 4.2 kHz, the depth and the diameter of the steel reinforcing bar can be evaluated. Using higher excitation frequency of about 80 kHz, and the X, Y signals of the lock-in amplifier, where the X signal is the same phase signal with the AC excitation magnetic field and Y signal is the 90° phase different signal with the AC excitation magnetic field, we could evaluate the corrosion of steel rebar. A compact system with low power consumption of 0.5 W was developed, and we also did some field experiments using this system.

Keywords: steel reinforcing bar, electromagnetic method, concrete, corrosion, eddy current, nondestructive evaluation

1. Introduction

The corrosion of steel reinforcing bar (rebar) in concrete structures not only reduces the strength of the concrete structures but also causes the breakage of the concrete due to the volume increase of the corrosion products. The corrosion of steel rebar is the leading cause of deterioration of concrete. The periodic inspection of the steel rebar in concrete is necessary and important. Knowing the conditions of the steel rebar in concrete structures, such as the location, the depth, the diameter, and the corrosion status of steel rebar, is important for the safety evaluation of concrete structures.

In Japan, many railway bridges were built 40 years ago. It is necessary to develop easy-operation method to evaluate the corrosion of steel rebar. A project about the corrosion evaluation of steel rebar was started in 2015. In this project, we have two objectives: one is to evaluate the depth and the diameter of steel rebar in concrete; another is to evaluate the corrosion of steel rebar in concrete.

Ultrasonic method [1, 2], X-ray method [3, 4], magnetic flux leakage (DC field) method [5–7], electromagnetic induction (AC field) method [8–13], microwave radar system [14–16], and thermography technology [17–18] have been used to evaluate the break, the location, or the corrosion of the steel rebar in concrete.

Ultrasonic methods are often used to measure the crack of the concrete and the delamination between steel rebar and concrete. The advantage of ultrasonic method is its big detection depth till to several meters. The disadvantage of ultrasonic method is the noise caused by the reflection of stones in concrete.

Concrete scanning and imaging with X-ray can be used to get the position, diameter, and corrosion information of steel rebar in concrete. The advantage is its high spatial resolution. The disadvantage is its complex and high price and unsuitability for field experiments.

The magnetic flux leakage (MFL) method is mainly used to detect breaks in prestressing steel of pretensioned and post-tensioned concrete structures. The microwave radar systems can be used to detect the position and the covering depth of the steel rebar; however, the water or moisture in the concrete structures may influence the detection accuracy of the detection, and it is difficult to detect the second-layer steel rebar for the concrete with steel rebar grid. Thermography technology is not suitable to detect the steel rebar with deep depth.

Compared with other methods, the low-frequency electromagnetic induction method has the advantages of low cost and easy operation. It can be used to detect the covering depth and diameter of the steel rebar. For the low-frequency electromagnetic method, the moisture of concrete also has less influence to the detection results. However, the depth resolution or depth accuracy of commercial electromagnetic system is not good, which is difficult to be used to evaluate the diameter reduction of steel rebar due to corrosion.

We developed electromagnetic evaluation methods to detect the depth, the diameter, and the corrosion of steel rebar.

In this chapter, we will summarize our researches on the nondestructive evaluation of steel rebar. Firstly, we will give an introduction of the corrosion mechanism and the corrosion products of the steel rebar in concretes; secondly, we will describe the setup and results of depth and diameter measurements using dual-frequency and dual-probe method; thirdly, we will discuss the experiments and results of corrosion evaluation of steel rebar; and finally, we will provide the conclusion.

2. Corrosion mechanism, corrosion products, and electromagnetic method

The corrosion of steel rebar in concrete is complex, but basically it is an electrochemical reaction similar to that of a simple battery. **Figure 1** shows the corrosion mechanism of steel rebar in concrete with seawater environment. The corrosion of steel rebar has strong relation with the PH value of the environment in concretes. The corrosion of the steel reinforcement occurs below pH 11. Initially, on the surface of steel rebar, a protective oxide coating is formed giving a pH of 13–14 adjacent to the steel. When the carbonation and chloride ions diffuse through the concrete and reach the embedded steel rebar, then the pH is lowered by the presence of the chloride ions. This corrosion process also requires oxygen diffused through the concrete, setting up an electrochemical reaction. The process is increased by the existence of voids in the concrete adjacent to steel.

As the corrosion process progresses, corrosion products tend to build up in certain areas of the steel rebar. These corrosion products have different elemental compositions than the original state.

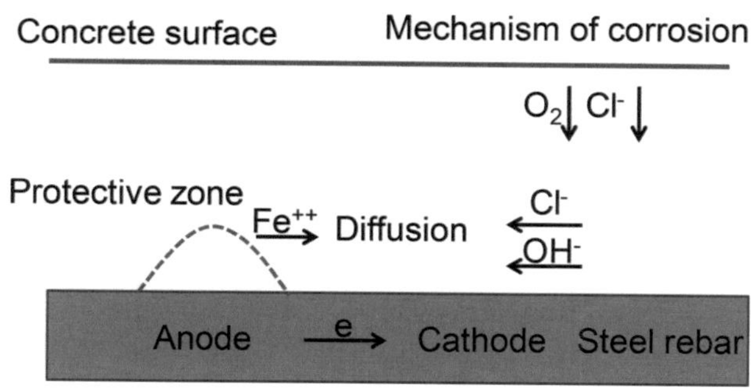

Figure 1.
Corrosion mechanism of steel rebar in concrete.

	Relative permeability	Conductivity (S/m)
Steel rebar	100	7×10^4
$FeO{\cdot}nH_2O$ ($Fe(OH)_2$)	1.0072	300
FeOOH (α, β, γ, δ)	1.1	5×10^{-9}
$Fe_2O_3{\cdot}H_2O$ ($Fe(OH)_3$)	Hundreds (soft)	10^{-4}
Fe_3O_4	Hundreds (hard)	200

Table 1.
Corrosion products and their electrical conductivities and permeabilities.

Almeida et al. [19] reported that the initial stages of atmospheric corrosion of carbon steel in both rural and urban atmospheres yield the formation of lepidocrocite (γ-FeOOH) and goethite (α-FeOOH). In chloride-containing environments, another form of iron oxyhydroxide, akaganeite (β-FeOOH), is frequently found [20]. Magnetite (Fe_3O_4) is also frequently reported in coastal environments [21]. Lepidocrocite and goethite are reported as the main phases independent of the environment [22]. Raman spectroscopy has been successfully used to characterize the different oxide phases formed on iron-based alloys [23]. What we normally call rust is a flaky red-brown solid which is largely hydrated iron. The primary corrosion product of iron is $Fe(OH)_2$ (or more likely $FeO{\cdot}nH_2O$), but the action of oxygen and water can yield other products having different colors: $Fe_2O_3{\cdot}H_2O$ (hydrous ferrous oxide, sometimes written as $Fe(OH)_3$) is the principal component of red-brown rust. It can form a mineral called hematite. $Fe_3O_4{\cdot}H_2O$ ("hydrated magnetite" or ferrous ferrite, $Fe_2O_3{\cdot}FeO$) is most often green but can be deep blue in the presence of organic complexants as shown here. Fe_3O_4 ("magnetite") is black.

These corrosion products are on the surface of the steel rebar and have different electrical conductivities and permeabilities. **Table 1** shows several kinds of corrosion products and their electrical conductivities and permeabilities.

Steel rebar has high permeability and high electrical conductivity. All the corrosion products have lower electrical conductivity than steel rebar. For some corrosion products, such as $FeO{\cdot}nH_2O$ and FeOOH, the permeabilities become smaller, and the relative permeabilities are close to 1; for some corrosion products, such as $Fe_2O_3{\cdot}H_2O$ and Fe_3O_4, the permeabilities become bigger. $Fe_2O_3{\cdot}H_2O$ is a kind of soft magnetic material, and Fe_3O_4 is a kind of hard magnetic material. Because of the different electrical conductivity and permeability properties, the electromagnetic responses of these corrosion products are different. Therefore, it is possible to evaluate the corrosion of steel rebar using electromagnetic method.

Figure 2 shows the principle of electromagnetic method to evaluate the steel rebar. AC magnetic field is produced by the excitation coil when AC current flows

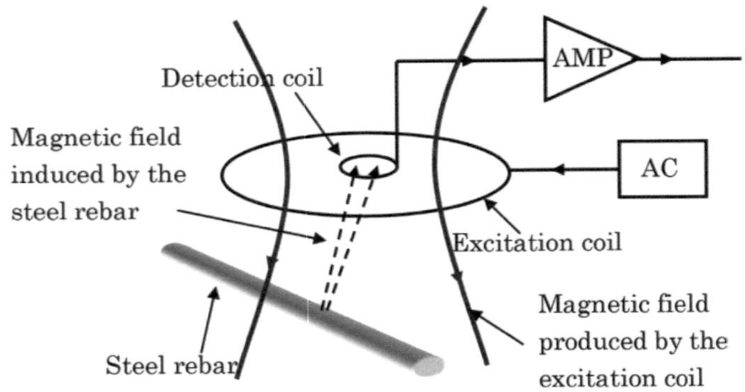

Figure 2.
Principle of electromagnetic method.

in it. Then, eddy current and magnetization are caused in the steel rebar. The detection coil is used to measure the magnetic field produced by the eddy current and the magnetization of steel rebar. The penetration depth of the eddy current has relation with the frequency, the electrical conductivity, and permeability of the material. It can be expressed by this formula:

$$\delta = \frac{1}{\sqrt{\pi f \mu \sigma}} \tag{1}$$

where δ is the penetration depth, f is the frequency, μ is the permeability, and σ is the electrical conductivity. When low-frequency electromagnetic field is used, the penetration depth of eddy current is big, and the magnetization effect which determines the signal, depth, and diameter of steel rebar can be evaluated. When high-frequency electromagnetic field is used, the penetration depth is small; mainly, the surface property of steel rebar determines the signals; thus the corrosion of steel rebar can be evaluated. For steel rebar, the relative permeability μ_r is about 100. The electrical conductivity is about 7×10^6 S/m. Using Formula (1), we can calculate the penetration depth. It is about 0.32 mm when the frequency is 4 kHz, and it is about 0.07 mm when the frequency is 80 kHz.

We find the magnetization effect has more contribution to the signal at 4 kHz, so we will use the frequency of about 4 kHz to evaluate the depth and diameter of steel rebar. And the eddy current effect has more contribution to the signal, and the penetration depth is small at 80 kHz, so we will use the frequency of about 80 kHz to evaluate the corrosion of steel rebar.

3. Depth and diameter detection of steel rebar

To measure the depth and the diameter of steel rebar simultaneously, two probes and dual-frequency method were used. **Figure 3** shows the setup of the measuring system.

For each probe, it was composed by a signal generator, an excitation coil, and a detection coil. The signal generators were used to produce the AC current and sent to the excitation coils to produce the AC magnetic fields. The magnetic fields induced by the steel rebar due to the eddy current and the magnetization were detected by the detection coils. The output signals of the detection coils were amplified by the amplifiers. The demodulators were used to get the amplitudes of

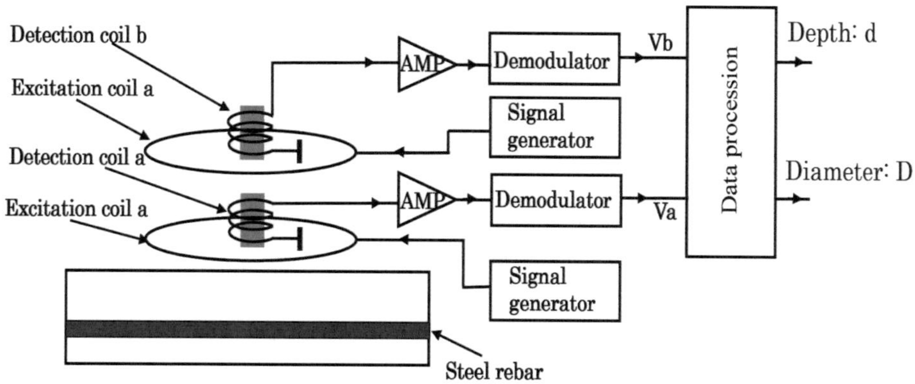

Figure 3.
Experimental setup for the depth and diameter evaluation of steel rebar.

Depth (mm)	Diameter (20 mm)	Diameter (16 mm)	Diameter (13 mm)	Diameter (10 mm)
20	1.03142	0.85424	0.61139	0.37896
30	0.28668	0.23217	0.1742	0.10764
40	0.1011	0.08372	0.06353	0.04016
50	0.04279	0.03493	0.02726	0.01713
60	0.02003	0.01641	0.01276	0.00846
70	0.0101	0.00839	0.00651	0.00419
80	0.00555	0.00449	0.0036	0.0023
90	0.00312	0.0026	0.00201	0.0014
100	0.00188	0.00155	0.00133	0.00095

Table 2.
Signal amplitudes of V_a of probe a for the steel rebars with different diameters and depths.

the signals. After the data processing and calculation, the covering depth and the diameter of steel rebar were calculated.

The experimental conditions were as follows: the diameters of the two excitation coils were both 70 mm with the turns of 100. The two detection coils were both 200 turns with the diameter of about 10 mm. The gains of the amplifiers were 20 dB. The distance between the two detection coils was about 10 mm. The current amplitudes of the AC currents flowing in the excitation coils were both about 20 mA. To reduce the interference between the two probes, different frequencies were used for the two probes. The frequency of signal generators were 3.8 and 4.2 kHz, respectively, and low-pass filters were used after the demodulators.

For the steel rebar with different depths and different diameters, the amplitudes V_a and V_b of the output signals of the two probes were different. **Table 2** shows the signal amplitudes of V_a for the steel rebars with different diameters of 10, 13, 16, and 20 mm and depths from 20 to 100 mm. **Table 3** shows the signal amplitudes of V_b for the steel rebars with different diameters and depths.

Figure 4a and **b** shows the signal amplitudes V_a and V_b changed with the depths for different steel rebar diameters of 10, 13, 16, and 20 mm.

Figure 5a and **b** shows the signal amplitudes of V_a and V_b changed with the diameter of steel rebar with the depth of 30 mm. We can see the signals increased almost linearly with the diameter of the steel rebar. For other depths, the results were similar.

Depth (mm)	Diameter (20 mm)	Diameter (16 mm)	Diameter (13 mm)	Diameter (10 mm)
20	0.68276	0.55748	0.40017	0.25621
30	0.23165	0.18534	0.13871	0.08838
40	0.09557	0.07794	0.05806	0.03805
50	0.04572	0.03646	0.02731	0.01843
60	0.02359	0.01877	0.01402	0.0093
70	0.01261	0.01025	0.00731	0.00531
80	0.00723	0.00589	0.00447	0.00314
90	0.00453	0.00352	0.00253	0.00192
100	0.00301	0.00219	0.00162	0.00123

Table 3.
Signal amplitudes of V_b of probe b for the steel rebars with different diameters and depths.

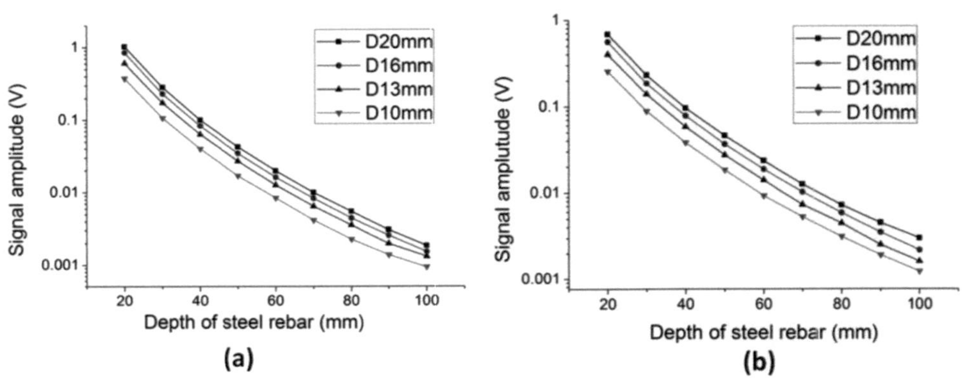

Figure 4.
(a) For different diameters of steel rebar, the signal amplitude of V_a changed with the depth of steel rebar. (b) For different diameters of steel rebar, the signal amplitude of V_b changed with the depth of steel rebar.

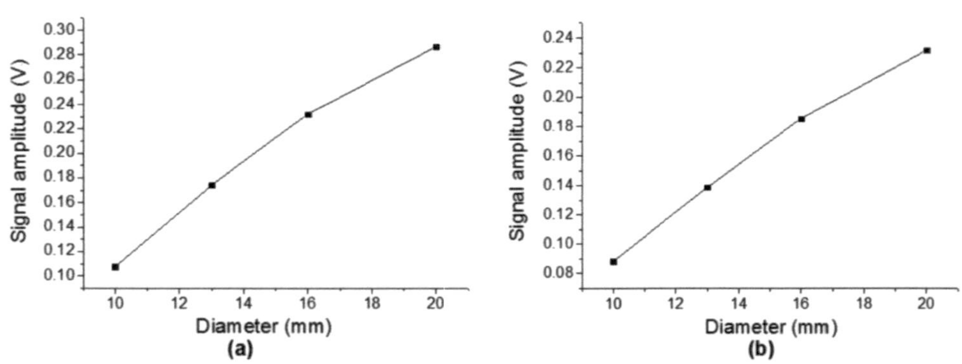

Figure 5.
(a) Signal amplitudes of V_a changed with the diameter of steel rebar with the depth of 30 mm. (b) Signal amplitudes of V_b changed with the diameter of steel rebar with the depth of 30 mm.

According to the data of **Tables 1** and **2**, a fitting formula was given to express the signal amplitudes changing with the diameter and the depth of steel rebar. The signal amplitude V_a of the lower probe was expressed by Formula (2):

$$V_a = kD \cdot e^{\left(-\beta d + \frac{\gamma}{\delta + d}\right)} \tag{2}$$

If the distance between the lower probe and the upper probe was L, the signal amplitude V_b of the upper probe was expressed by Formula (3):

$$V_b = kD \cdot e^{\left(-\beta(d+L) + \frac{\gamma}{\delta + d + L}\right)} \tag{3}$$

where V_a was the output signal amplitude of the lower probe and V_b was the output signal amplitude of the upper probe. D was the diameter of steel rebar, and d was the covering depth of steel rebar. L was the distance between the lower probe and the upper probe. α, β, and γ were constants.

Using Formulaa (2) and (3), we can get:

$$\frac{V_a}{V_b} = e^{\beta L} \cdot e^{\left(\frac{\gamma}{\delta + d} - \frac{\gamma}{\delta + d + L}\right)} = e^{\beta d_0} \cdot e^{\frac{\gamma L}{(\delta + d) \cdot (\delta + d + L)}} \tag{4}$$

From Formula (4), V_a/V_b was mainly determined by the covering depth of steel rebar, and it had less relation with the diameter of steel rebar.

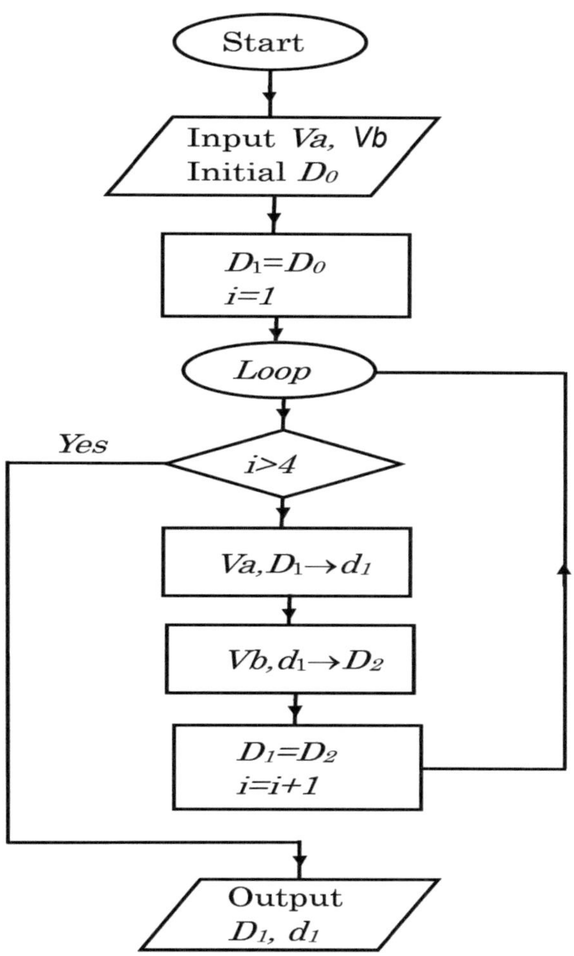

Figure 6.
Program flowchart of the recursion calculation method to get the depth and the diameter of the steel rebar.

For a steel rebar with unknown depth and diameter, V_a/V_b was obtained using the measured values of V_a and V_b. The covering depth d can be calculated using Formula (4), and the diameter D can be calculated using Formula (2) and the calculated depth d.

We can also calculate the depth d and the diameter D of the steel rebar using a recursion calculation method. **Figure 6** shows the program flowchart of the recursion calculation method. In the chart, D means the diameter of the steel rebar, and d means the depth of the steel rebar. First, set the diameter D_1 equal to the initial value of D_0. Then, use V_a and D_1 to calculate the depth of d_1, use V_a and d_1 to calculate D_2, and then repeat the calculation. The recursion calculation ended after calculating four times. Finally output the values of depth d_1 and the diameter D_1.

The result using the recursion calculation method and the calculation result using Formulas (1)–(3) were similar.

Table 4 shows the real values and the measured values of the depth and diameter of steel rebar using the recursion calculation method. For the steel rebar with depth of about 50 mm, the error of the depth measurement was about 0.5 mm, and the error of the diameter measurement was about 1 mm.

Real diameter (mm)	Real depth (mm)	Measured diameter (mm)	Measured depth (mm)
10	20	10.73	20.45
	30	10.30	30.35
	40	09.73	39.60
	50	09.65	49.50
	60	09.02	58.20
13	17	13.35	17.33
	27	12:48	28.83
	37	12.01	36.55
	47	12.16	46.58
	57	12.14	56.50
	67	11.50	65.45
16	14	15.11	14.24
	24	15.35	23.60
	34	15.12	33.65
	44	16.80	44.45
	54	15.05	53.60
	64	16.63	62.97
20	10	19.91	10.55
	20	20.67	19.85
	30	19.67	29.82
	40	20.09	39.80
	50	19.05	49.50
	60	19.24	60.75

Table 4.
Real values and measured values of the depth and diameter of steel rebar.

4. Corrosion evaluation of steel rebar

When corrosion happened with the steel rebar in concrete, corrosion products were produced on the surface of the steel rebar. The corrosion products had different electrical conductivity and permeability, so the electromagnetic response of the corrosion products was different from that of the steel rebar. Therefore, it was possible to evaluate the corrosion of steel rebar by measuring the electromagnetic response. **Figure 7** shows the experimental setup for the corrosion evaluation of steel rebar using electromagnetic method.

The excitation coil was used to produce the AC magnetic field, and eddy current was induced in the steel rebar. The detection coil was used to measure the magnetic field produced by the steel rebar. The lock-in amplifier was used to get the X and Y signal, where the X signal was the same phase signal with the excitation magnetic field and the Y signal was the 90° phase different signal with the excitation magnetic field.

Figure 8 shows the compact system of corrosion evaluation of steel rebar. The excitation coil, the detection coil, the amplifier, and the lock-in amplifier were integrated in a small box of about 9 × 12 × 7 cm. Only one USB cable was used to

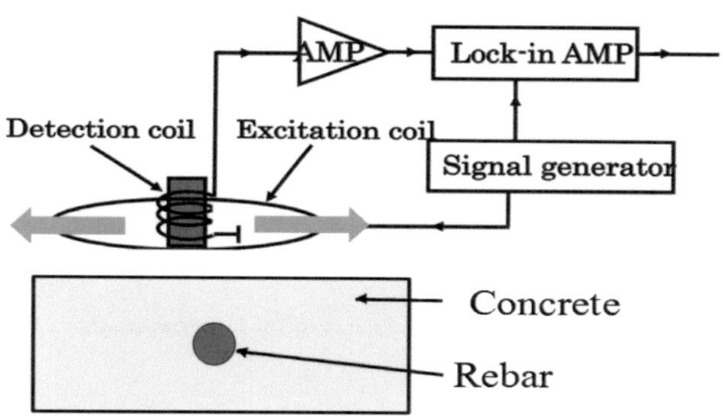

Figure 7.
Experimental setup for the corrosion evaluation of steel rebar using electromagnetic method.

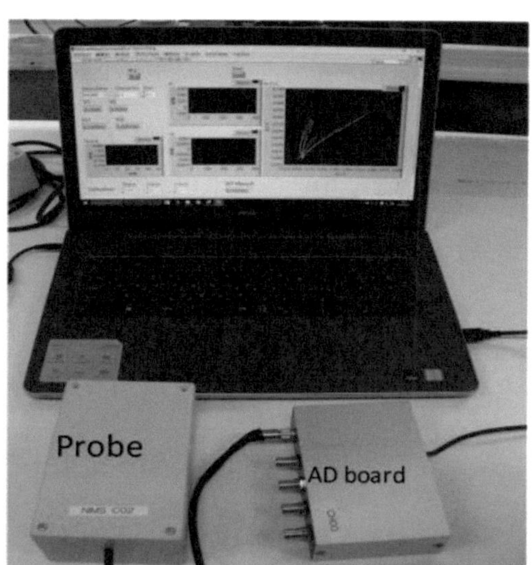

Figure 8.
Compact system for the corrosion evaluation of steel rebar.

connect it with a computer. The power of the system was also from the USB cable. The total power consumption of the system was about 0.5 W.

Figure 9 shows the samples of steel rebar with different corrosion levels. The diameter of the steel rebar was 16 mm. The steel rebar "*a*" had no corrosion; steel rebar "*b*" had a little corrosion, and there are some corroded dots on the surface the steel rebar; steel rebar "*c*" had big corrosion, and the thickness of the corroded layer was about 0.1 mm; and steel rebar "*d*" had severe corrosion with the thickness of the corrosion layer of about 1 mm.

In our experiments, the diameter of the excitation coil was 3 cm with 100 turns. The diameter of the detection was 1 cm with 100 turns. The frequency was 80 kHz and the current amplitude was 20 mA. We scanned the steel rebars using the system. **Figure 10** shows the signals of steel rebar with different depths of 2, 3, 4, and 5 cm.

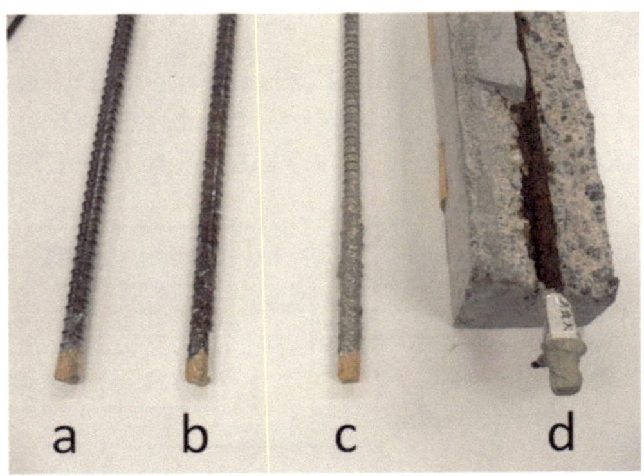

Figure 9.
Samples of steel rebar. (a) No corrosion, (b) a little corrosion, (c) big corrosion, and (d) severe corrosion.

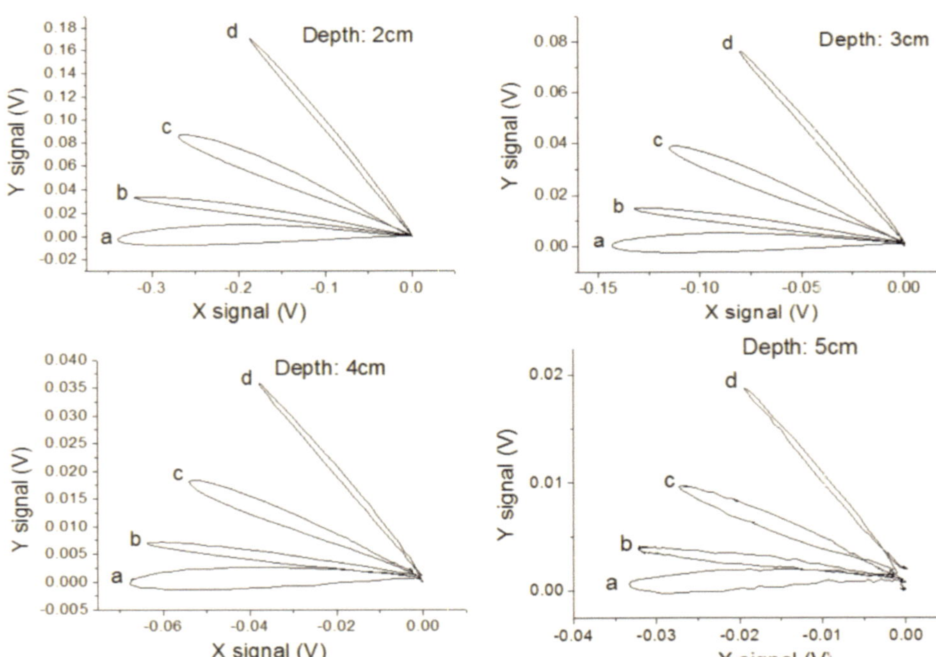

Figure 10.
Signals of steel rebars of (a), (b), (c), and (d) with different depths of 2, 3, 4, and 5 cm.

X-Y graphs were plotted using the X and Y output signals of the lock-in amplifier. The depth only changed the signal amplitudes of the X and Y signals, and it almost had no influence to the slope of the X-Y graph. For different corrosion levels, the slopes were different. From the slope, the corrosion of steel rebar can be judged.

The absolute value of ΔY/ΔX increased with the corrosion level. **Figure 11** shows the absolute values of ΔY/ΔX for the steel rebars at different depths. They were mainly determined by the corrosion levels and had less relation with the covering depth. This result proved that it was possible to evaluate the corrosion of steel rebar using this electromagnetic method.

We also did some field experiments using our system. The depths of steel rebars were measured. **Figure 12** shows it. The experiments were done under a railway bridge. When train passed over the bridge, there was almost no influence to our experiments. We did the scanning by hand. After measurements, the concrete was removed, and the real depths of steel rebars were measured. Then compare the real values with the measured values using electromagnetic method.

The steel bars were crossed into a mesh. The distance between the steel rebar was about 20 cm. **Figure 13** shows the real values and the measured values of steel rebar. The numbers in parentheses were measured values. For some rebars, the measured depths were consistent well with the real depths. For some rebars, there were big differences between the real values and the measured values. We found the difference was small, and the steel rebar had almost no corrosion; when the difference was big, the steel rebar had corrosion. When the steel rebar was corroded, the electrical conductivity and the permeability changed, which caused the difference between the real value and the measured value.

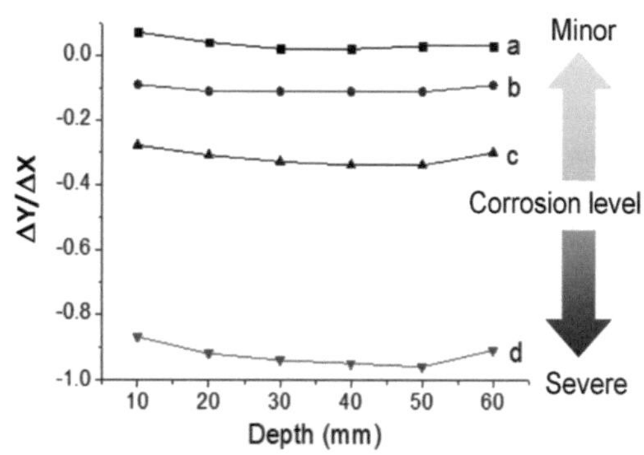

Figure 11.
Values of ΔY/ΔX for the steel rebars at different depths.

Figure 12.
Field experiments of checking the railway bridge.

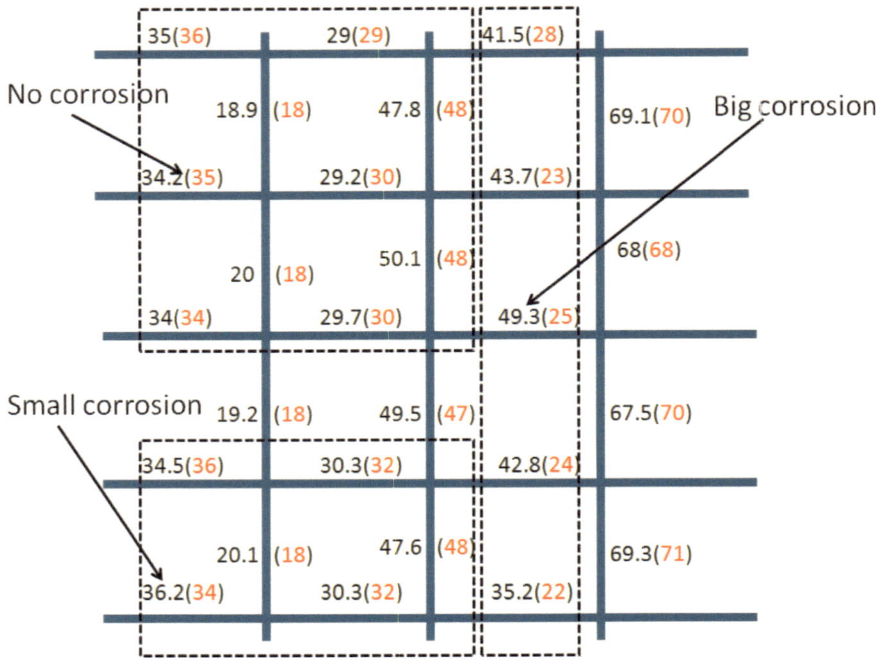

Figure 13.
Real values and the measured values of steel rebar. The numbers in parentheses were measured values. When the difference was big, the corrosion of steel rebar was also big.

5. Conclusion

Knowing the conditions of the steel reinforcing bar (steel rebar), such as the location, the diameter, and the corrosion of steel rebar, is important for the safety evaluation of concrete structures. We developed an electromagnetic method to detect the depth, the diameter, and the corrosion of steel rebar in concrete.

Using the lower frequencies of about 3.8 and 4.2 kHz and two probes, the depth and the diameter of steel rebar could be evaluated simultaneously. The resolution of about 1 mm was achieved.

Using higher frequency of about 80 kHz, the corrosion of steel rebar could be evaluated. A compact system with a low power of about 0.5 W was developed, and only one USB cable was used for the power and data transferring. The excitation coil, the detection coil, the amplifier, and the lock-in amplifier were integrated in a small box of about 9 × 12 × 7 cm. Using the X signal and Y signal outputs of the lock-in amplifier, where the X signal is the same phase signal with the excitation current and the Y signal is 90° phase difference signal with the excitation current, we plotted X-Y graph of steel rebar signals. From the slope of the X-Y graph, the corrosion of steel rebar could be evaluated.

We also did some field experiments using our system. The depths of the steel rebar in a concrete railway bridge were evaluated. If the steel rebar had no corrosion, the measured value is consistent well with the real value. If the steel rebar had big corrosion, the difference between the real value and measured value was also big. The reason is that the corrosion products have different electromagnetic properties with the steel rebar. We will do more field experiments and improve our system in the future.

Acknowledgements

This work was supported by the Council for Science, Technology and Innovation, “Cross-ministerial Strategic Innovation Promotion Program (SIP), Infrastructure Maintenance, Renovation, and Management,” (Funding agency: JST). We thank the helps from Dr. M. Shiwa, Dr. S. Takaya, Mr. S. Aramaki, Mr. N. Tsutsumi1, and Dr. K. Tsuchiya.

Author details

Dongfeng He
National Institute for Materials Science, Tsukuba, Japan

*Address all correspondence to: he.dongfeng@nims.go.jp

References

[1] Rucka M. Failure monitoring and condition assessment of steel-concrete adhesive connection using ultrasonic waves. Applied Sciences-Basel. 2018;**8**: article number 320

[2] Saleem M. Study to detect bond degradation in reinforced concrete beams using ultrasonic pulse velocity test method. Structural Engineering and Machanics. 2017;**64**:427-436

[3] Ostuka K, Takeda T. Research on internal shape detection of cracks in concrete structures. In: Proceedings of the 52nd JSCE Annual Meeting. Vol. V. 1997. pp. 692-693

[4] Beck M, Goebbels J, Burkert A, Isecke B, Babler R. Monitoring of corrosion processes in chloride contaminated mortar by electrochemical measurement and X-ray tomography. Materials and Corrosion. 2010;**61**:475-479

[5] Ghorbanpoor A. Magnetic-based NDE of steel in prestressed and post-tensioned concrete bridges. In: Proceedings Structural Materials Technology III. San Antonio, Texas; 1998. pp. 343-349

[6] Thomas W, Thomas V. Detection of reinforcement breaks on large-scale fatigue tests with the magnetic flux leakage method. In: 8th fib PhD Symposium in Kgs. Lyngby, Denmark; 2010. pp. 587-592

[7] Perin D, Goktepe M. Inspection of rebars in concrete blocks. International Journal of Applied Electromagnetics and Mechanics. 2012;**38**:65-78

[8] Gaydecki PA, Burdekin FM. An inductive scanning system for two dimensional imaging of reinforcing components in concrete structures. Measurement Science and Technology. 1994;**5**:1272-1280

[9] Yu ZZ, Gaydecki PA, Silva I, Fernandes BT, Burdekin FM. Magnetic field imaging of steel reinforcing bars in concrete using portable scanning systems. Review of Progress in Quantitative Nondestructive Evaluation. 1999;**18**:2145-2152

[10] Gaydecki P, Silva I, Fernandes BT, Yu ZZ. A portable inductive scanning system for imaging steel-reinforcing bars embedded within concrete. Sensors and Actuators A. 2000;**84**:25-32

[11] Miller G, Gaydecki P, Quek S, Fernandes B, Zaid M. A combined Q and heterodyne sensor incorporating real-time DSP for reinforcement imaging, corrosion detection and material characterization. Sensors and Actuators, A: Physical. 2005;**121**:339-346

[12] Ricken W, Mehlhorn G, Becker W. Determining of the concrete cover thickness and the bar diameter in reinforced concrete with a method of eddy current testing. In: Proceeding International Symposium Non-Destructive Testing in Civil Engineering. Berlin, Germany; 1995. pp. 197-204

[13] Miller G, Gaydecki P, Quek S, Fernandes BT, Zaid MAM. Detection and imaging of surface corrosion on steel reinforcing bars using a phase-sensitive inductive sensor intended for use with concrete. NDT & E International. 2003;**36**:19-26

[14] Bungey JH, Millard SG. Radar inspection of structures. In: Proceedings of the Institution of Civil Engineers—Structures and Buildings. Vol. 99. 1993. pp. 173-178

[15] Molyneaux TCK, Millard SG, Bungey JH, Zhou JQ. Radar assessment of structural concrete using neural networks. NDT & E International. 1995; **28**:281-288

[16] Shaw MR, Millard SG, Molyneaux TCK, Taylor MJ, Bungey JH. Location of steel reinforcement in concrete using ground penetrating radar and neural networks. NDT & E International. 2005; **38**:203-212

[17] Keo SA, Brachelet F, Breaban F, Defer D. Steel detection in reinforced concrete wall by microwave infrared thermography. NDT & E International. 2014;**62**:172-177

[18] He Y, Tian G, Pan M, Chen D. Eddy current pulsed phase thermography and feature extraction. Applied Physics Letters. 2013;**103**:084104

[19] Almeida E, Morcillo M, Rosales B, Marrocos M. Atmospheric corrosion of mild steel part I—Rural and urban atmospheres. Materials and Corrosion. 2000;**51**(12):859-864

[20] Ma Y, Li Y, Wang F. Corrosion of low carbon steel in atmospheric environments of different chloride content. Corrosion Science. 2009;**51**(5): 997-1006

[21] Syed S. Atmospheric corrosion of hot and cold rolled carbon steel under field exposure in Saudi Arabia. Corrosion Science. 2008;**50**(6): 1779-1784

[22] Syed S. Atmospheric corrosion of carbon steel at marine sites in Saudi Arabia. Materials and Corrosion. 2010; **61**(3):238-244

[23] Hazan E, Sadia Y, Gelbstein Y. Characterization of AISI 4340 corrosion products using Raman spectroscopy. Corrosion Science. 2013;**74**:414-418